THE QUANTUM GUIDE TO LIFE

THE QUANTUM GUIDE TO LIFE

How the Laws of Physics Explain Our Lives from Laziness to Love

KUNAL K. DAS

Skyhorse Publishing

Skyhorse Publishing books may be purchased in bulk at special discounts for sales promotion, corporate gifts, fund-raising, or educational purposes. Special editions can also be created to specifications. For details, contact the Special Sales Department, Skyhorse Publishing, 307 West 36th Street, 11th Floor, New York, NY 10018 or info@skyhorsepublishing.com.

www.skyhorsepublishing.com

10 9 8 7 6 5 4 3 2 1

Library of Congress Cataloging-in-Publication Data available on file

ISBN 978-1-62087-624-4

Printed in the United States of America

CONTENTS

ACKNOWLEDGMENTS

Writing this book was mostly a solitary effort channeling ideas that I have formed over several years. I am responsible for all the ideas, the text, and the figures in this book. But after having written (and drawn) most of it for fun, when I thought of publishing it, I discovered that finding a publisher as an unknown, first-time author can be even more challenging than writing. That process was successful largely because I found the perfect literary agent for this book, Andy Ross, who shared my excitement about it. As the only person other than me who read all the chapters in detail before submission, his feedback was extremely valuable. My sincere thanks also go to my editor, Jennifer McCartney, at Skyhorse Publishing, for appreciating the value of a rather unconventional book. Her help and advice has been invaluable in transforming the manuscript into a published book. I am grateful to Rosario Torres, Sudarshan Fernando, and Eric Laub for helpful comments on some of the chapters. I owe much to my parents for raising me in an environment where I was always free to choose my own path and own beliefs in life and for encouraging diverse interests from early on.

INTRODUCTION

Ask your friends, "What are your thoughts on life?" "How do you explain human behavior?" "Why are relationships so hard?" or questions of that sort, and you're sure to get all kinds of answers, some obvious, some funny, some wild and whimsical, but I bet none of them would involve the laws of quantum physics. After all, what could quantum physics possibly tell us about our lives or the people around us? A great deal, actually, as this book will show. In quantum physics, we can find answers and justifications for almost every aspect of our lives, from the fundamentally profound—such as why life is so full of compromises—to the utterly mundane—like why it always takes effort to keep things tidy. As we seek out those hidden quantum connections and parallels, we will come up with often surprising and sometimes amusing insights into our life experiences. The universality of the themes considered here means that this book is truly intended for all literate practitioners of this thing we call life. Therefore, to make sure that no one is left out, I deliberately avoid math and dense logic with the firm belief that they are not necessary to communicate the essence of great ideas, even in quantum physics.

Two realms, which have always been considered antipodal and absolutely *un*-relatable, come together in this book: the realm of everyday life (personal, social, relational, and financial) *and* the realm of the fundamental laws of physics. Therefore, you can read this *either* as a book about life that offers a completely different and novel perspective *or* as a brand new kind of popular science book where the abstract concepts of quantum physics are framed in terms to which we can all relate.

Although the book is intended for a broad audience, physicists will also find much of interest here. As far as I know, this is the only serious attempt ever at bridging what I call the "Physicists' Dilemma," which is that, while they themselves are absolutely convinced that they are dealing with the most important things in the universe, they can rarely make their work seem relevant or genuinely engaging, even to their friends and families. Fascination and curiosity are not enough—*relevance* is crucial. The goal here is to make quantum physics relevant and relatable to everyone.

Each chapter can be read independently of all the other chapters. So, if you get a bit stuck on some idea in one chapter, feel free to move on to another one and come back to it. Every chapter explores the implications of one idea or set of closely related ideas important in quantum physics. But with physical laws being generally interconnected, there are bound to be occasional references to other chapters. However, in most cases, they are not absolutely essential to understanding the material at hand, and I will point out the few exceptions as and when they arise.

The ideas presented here draw intriguing parallels between the rules of life and the laws of quantum physics. This correspondence is relevant, because almost all our knowledge and wisdom is based on our drawing analogies and connections between what we learn *and* what we already know well. Often, the parallels are so eerily close as to make us wonder if our consciousness really sets us that much apart from the physical and inanimate universe, beyond the reach of its laws.

CHAPTER 1
A QUANTUM OF HAPPINESS

Quantization happens when a physical system is restricted by boundary conditions, to exist in only certain very specific states. Likewise, our mental states, as in how we feel, depend on the boundary conditions in our lives, meaning all the bonds and constraints that keep us where we are, with whom we are, and doing what we do. The mechanisms that can alter quantum states suggest ways to reach happier states in life.

"Money cannot buy you happiness"—an ancient bit of wisdom, often quoted, but seldom believed. You must have heard it often enough and perhaps even mutter it yourself occasionally. But if you do not feel wealthy, I bet you always had a lurking suspicion that this can't be right—surely you would be a lot happier if you had more money! Rich folks do seem to be generally happier, and wealth certainly opens up a lot of attractive possibilities. So it is only natural to doubt and wonder, particularly in the ultra-materialistic world we live in today where we have become absolutely dependent on our possessions. We often tend to feel that the more we possess, the better we would feel, and the happier our lives would be. Shopping has indeed become a wistful antidote for feeling down and low.

Life does not come with a guarantee of happiness. But that has never prevented anyone from feeling entitled to it, and we stay in hard pursuit of it all our lives. Ironically, by trying so hard, we often make it even more elusive. In fact, we can't even agree upon what happiness really means. Ask around: You will simply get personalized descriptions of everything happiness is not—laundry lists of all those things missing in people's lives that they think are keeping them from being happy. Each list would be a bit different, indicating a different definition of happiness for everyone.

So, with no universal definition or prescription on how to find it, our collective quest for happiness continues to be an essentially blind quest, and like any blind quest, the failure rate is quite high. Wouldn't it be nice to get an objective perspective on what happiness is all about—to be able to establish a few concrete facts that could guide us in this universal quest for happiness? With some almost poetic parallels, the character of quantum states can help us do just that.

Let us start with the one thing we can be sure of: Happiness is a state of mind. Although a bit of a cliché, this never fails to impress anyone stumbling upon it for the first time as a profound bit of insight. And it is essentially true; whether we feel happy or sad, it's all just a state of mind. But it does not tell us much that is of any practical use, such as how we might be able to influence and change our mental states.

This is where quantum mechanics comes in handy because quantum mechanics is all about "states": eigenstates, position states, momentum states, closed states, open states, bound states, entangled states, stationary states—even the whole universe is speculated to be in a quantum state. In quantum mechanics, the *state* of a system is simply the status of all the characteristics that describe it.

Rather like how you might describe your own "state" right now—as a list of all the relevant variables in your life as they happen to be playing out currently. A subset of those variables that influence how you feel right now defines your current mental "state." Quantum mechanics has been rigorously dealing with all kinds of *states* for a whole lot of complicated things, so it can certainly give us a few pointers about the states of the mind as well. After all, each human being, and therefore the human mind, is defined by some sort of quantum state as well, albeit a very complex one. In recent years, there actually have been some serious attempts by respected scientists to explain consciousness with quantum theory.

The most important class of states in quantum mechanics are the *stationary* states, because they really got the whole field started. And the name says it all—once a system is in a stationary state it will remain there, stationary and unchanging, unless disturbed. Such states have some remarkable properties, as we will see. Introduced by the Danish physicist Niels Bohr in 1915, the concept of stationary states was the real birth of quantum mechanics, establishing it as a completely different view of nature, distinct from classical Newtonian physics.

It is particularly appropriate to begin our quest for happiness with stationary states, because in real-life terms, the stability of an unchanging stationary situation usually marks the first step toward happiness, implying that at least things are not getting any worse! Indeed, we all seek some level of stability in life to give us a sense of security. We have a name for the desirable stationary states in life: *states of contentment.* If you can manage to be content most of the time with what you have, then you can be reasonably assured that your life has been a success. As with quantum states, numerous stationary states are possible for every individual—each of those states corresponding to a different combination of situations with which someone could be content. Despite popular beliefs to the contrary, we don't all need to be rich and famous to be content. For example, you could be content with a stable desk job with a caring family to return to every evening in a middle-class suburb; or you could be enjoying the single life as a millionaire actor in a successful sitcom with a mansion by the beach and a Ferrari in your garage; or you could even achieve a high level of contentment working the night shift in the local four-year college where your kids can attend for free, and you enjoy your local community activities and the bowling league. We can keep adding to the list and could potentially find stability and contentment in any one of a wide variety of life situations. Yet, as we all know, lasting contentment is not easy to find, and that is because there is something very particular about stationary states—in life, as well as in quantum mechanics.

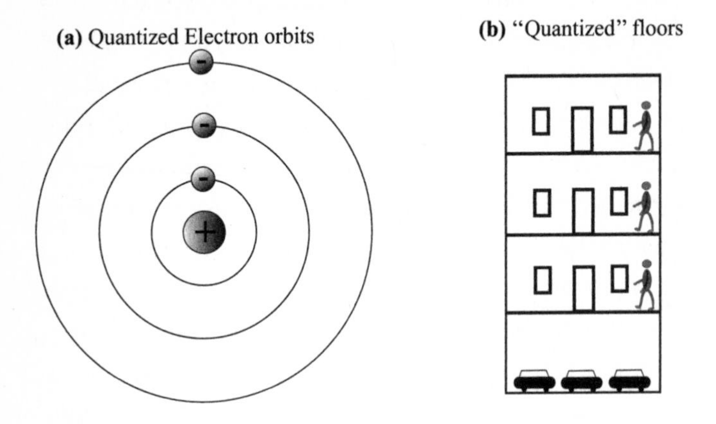

Figure 1.1 (a) Electron orbits around the nucleus of an atom are shown as concentric circles. The orbits are quantized, meaning only orbits with specific radii (distance from the center) are allowed. (b) This is analogous to the location of people living inside a multistoried building. People (like the electrons) can only be at specific elevations (like orbital radii) corresponding to the existing floors (except when they are in transit on the stairs).

After all, if stability were all there was to stationary states, Bohr would hardly have had to start a whole quantum revolution on account of them. You see, the most interesting thing about stationary states is that stationary states are very *specific*; we can't just pick any available state of the system and call it a stationary state. And the reason goes straight to the heart of what is *quantum* about quantum mechanics.

Although pretty much any quantum system can have stationary states, the clearest way to understand them is in terms of the states of an electron inside an atom. We all learn in school that every atom is like a little solar system, with a tiny compact nucleus made of protons (with positive electric charge) and neutrons (with no electric charge), with even tinier particles called electrons (with negative electric charge) in orbit around the nucleus just like the planets around the sun. However, there is a fundamental difference: In the solar system, the planets could in principle revolve around the sun at any radius or distance from the sun, so the earth could have been *arbitrarily* closer or farther than where it is now relative to the sun, and it could still have a perfectly stable orbit around the sun. But that is not the case with electrons. If we draw an atom as shown in Figure 1.1 with a nucleus at the center and a bunch of circles around it to represent electron orbits, then according to quantum theory, those circles could not be of just any radius; the electron orbits can have only certain fixed allowed radii. This means that in the figure, if the circles drawn correspond to the smallest three allowed orbits, then we cannot draw some other circles in between them to create some intermediate orbits. The situation is just like that for the floors in a multistoried building. Suppose each floor is ten feet high, then people can occupy rooms at ten, twenty, or thirty feet of elevation from the ground (assuming the ground floor is a garage), as shown in Figure 1.1, but nobody can be in a room fifteen feet above the ground, because there is no such floor. It is likewise with electrons in their orbits. Electrons in the allowed orbits are in their stationary states, and they would remain there forever, unless disturbed. This striking phenomenon where only specific orbits are allowed is called the *quantization* of electronic orbits, because the orbital radii can only take discrete or *quantized* values. The reason this quantization happens is rather surprising, as we will see at the end of this chapter.

This finicky nature of stationary states gives a quantum perspective on the elusive nature of long-lasting personal states of contentment. In our lives, even more so than with the relatively simple electrons, a lot of things have to be just right to achieve a stable and lasting situation that would make us content. Even the least demanding among us is unlikely to be in a perpetual state of contented bliss, under just any arbitrary set of circumstances. Things would be a lot easier if we were all that easy to please! Getting all the conditions just right almost never happens! But when it does so once in a while, some lucky ones can hold on to a stationary state of contentment for a long time—we see people like that occasionally and might envy them.

But the real trouble for most of us is that even contentment is not enough: If you are fine with being content, very good for you—most people unfortunately are not! The truth of the matter is we crave happiness, not contentment. People don't write books about "pursuit of contentment"; Hollywood would not make movies about that. Contentment lies on the path to happiness, but usually is not the same as being happy.

Happiness or sadness is really all about changes. This might come as a surprise after all this talk about stationary states. Nevertheless, it is true, because it is only when things change that we register any feelings at all. If you feel a bit skeptical about that, that's probably because when most of us think of "change," we envision only major changes in life. But, by change, I mean *any change*, because every little incident that happens in life has the potential for making us happier or sadder. When a change is positive, leading us to a better situation than we are currently in, we are happy, and when it is negative and things get worse, we end up being sadder and unhappy—and how happy or unhappy depends on just how big the change is.

Think about it: If absolutely nothing ever changed in your life, in the short term you would reach some sort of equilibrium where you are neither happy or sad, but eventually you would just be bored out of your mind. That is why very few people are ever completely content. We are driven by our feelings and our need to feel, and being content is more like an absence of strong feelings. Change needs to happen to trigger the sensations of happiness we seek.

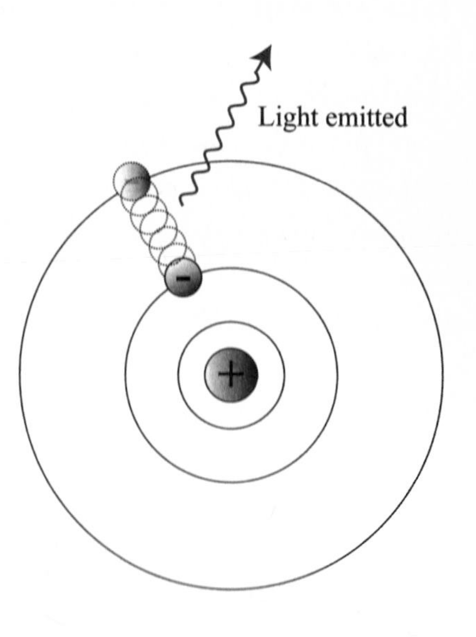

Figure 1.2 When an electron jumps from an outer orbit to an inner orbit, it loses energy, which is emitted as a particle of light (called *photon*). Vice versa, if the electron absorbs a photon, it gains energy to jump to an outer or higher orbit.

At a quantum level, changes happen all the time, but inside an atom, they happen in jumps. Since electrons only exist in very specific orbits, they cannot just ease into different orbits (there are no "stairs" among the different orbits like among the floors in a building); they have to jump to get from one orbit to another. And there is magic in those jumps—just as magical as true happiness.

Bohr did not get a Nobel Prize just for suggesting stationary states; he realized the wonderful thing that happens when electrons jump between stationary states: *light* happens! That's right, the tiny electrons dancing and jumping between stationary orbits is the origin of all the light in the universe. Here's how it works: An electron in an orbit with a larger radius has more energy than one in a smaller radius, and so when some perturbation triggers an electron to jump from an outer orbit to an inner one, the excess energy is released as a little packet or "quantum" of light, as shown in Figure 1.2. The reverse process can also happen: If a quantum of light with just the right amount of energy comes long, it can be absorbed by an electron to enable it to jump to an outer orbit. And just as the electron states are very specific, all the properties of the packets of light, so absorbed or emitted, are also very specific. Every jump between the same two energy levels will create clones of the exact same quantum of light, which, by the way, are called *photons* (hence photon torpedoes in *Star Trek*).[1]

We can visualize the changes in the state of our mind as happening similarly to the little quantum jumps of electrons inside an atom. Our mind remains in a stationary state until stimuli, external ones or internal ones (say due to bodily chemical shifts or memory flashbacks), lead to transitions in our mental state. At every waking moment of our life, there are things happening that influence our mood, with metaphorical quanta of happiness floating in and out: You could have been on your way to work at a job that you hate, and then the car radio confirms that you have won the lottery—that's a big quantum jolt of happiness—you go from being downright miserable to deliriously happy. Then there are the

[1] We never see individual packets of light floating around for the same reason that we do not see atoms and molecules. Light quanta, or *photons* as they are usually called, are very tiny and there are zillions of them in an ordinary beam of light. As a result, we see light as a continuum, the same way we see water as a fluid and never perceive the little molecules it is made of.

small quanta that change your mood a bit this way and that all the time: an attractive stranger smiled at you, and that made you a just a bit happier instantaneously. Someone behaved like a jerk for no good reason; your happiness drops a quantum. Most of the time, we simply receive too many stimuli on our mind and senses during our waking hours to distinguish individual "quanta" of happiness, so our change of mood might seem just as fluid as a beam of light composed of countless photons.

Viewed this way, perhaps it is not a coincidence that we have always associated light and brightness with happiness, and its absence and the descent into darkness with despair and gloom. The quantum analogy just reinforces all those metaphors we use to express feelings of joy: "Everything seems bright again," "The clouds are gone," "There is light at the end of the tunnel," "Every cloud has a silver lining," "If it is winter, can spring be afar?" (the sun fades in the winter; we anticipate its return to glory in the spring). And then all the ones for sadness: "The light is gone from my life," "It's all gone dark," "Why such a dark view of life?" The list goes on. Light is the absolute favorite metaphor and tool in literature, poetry, art, and movies to express the state of the mind, and as we now see, with some primordial roots in the very origin of light. Spread the quanta of happiness, spread the light!

Now let's get to the heart of the matter. What is it that defines the stationary states? Why is it that electron orbits can be of only specific radii? What is the reason for quantization? On the human side, what can we do to significantly change our stationary states of mind? To answer all that, we need to understand really what makes quantum mechanics, well . . . quantum.

The word *quantum* has become a cliché these days, used for all sorts of things, but few have a clear notion about what quantum really means and misconceptions abound. The word was coined by the German physicist, Max Planck, in 1901, when he suggested that the observed spectrum[2] of electromagnetic waves (which includes visible light, x-rays, ultra-violet rays, gamma rays, microwaves, and radiowaves) could be explained only by assuming that such waves (including ordinary visible light) actually come in discrete packets of energy that he called *quanta*. The idea was slow to catch on at first, but when it did, it caught fire and spurred intense research over the next three decades, which ushered in a completely new way of looking at the universe that has come to be known as *quantum mechanics*. The name underscores the fact that, as with light, many of the things in nature that were thought to exist as a continuum like a fluid actually come in discrete form like grains of sand. But there is a common misconception that everything in quantum mechanics is discrete or "quantized" and, vice versa, that discreteness is a unique feature of quantum physics. The discreteness is not so much about quantum mechanics *per se,* but is related to the fact that every system we deal with is finite and has boundaries. It is just that in the very small systems where quantum mechanics is most relevant, that discreteness is particularly conspicuous.

But how can boundaries make something discrete? It might seem obvious because all discrete or grainy little things have boundaries, due to their finite size. However, it is more subtle than that, because a river has boundaries, too, and we all think of water as a fluid. The way boundaries lead to discrete behavior in quantum mechanics is rather ironic, because to understand it, we need to look at waves, and waves essentially represent quite the opposite of discreteness—they are associated with continuous media like fluids. Therein lies a lingering mystery of the quantum world—the *wave–particle*

[2] Spectrum is like the fingerprint for different sources of electromagnetic waves. Everything in the universe radiates some form of electromagnetic waves (yes, even you—if you have ever seen thermal images of people, that is due to infrared waves emitted by the body). The spectrum of an object maps the intensities of various electromagnetic waves it emits. The spectrum changes with temperature—that is why we say "red hot" or "white hot," because as an object gets hotter, it not only emits heat (infrared) but visible light, as well, and starts to glow.

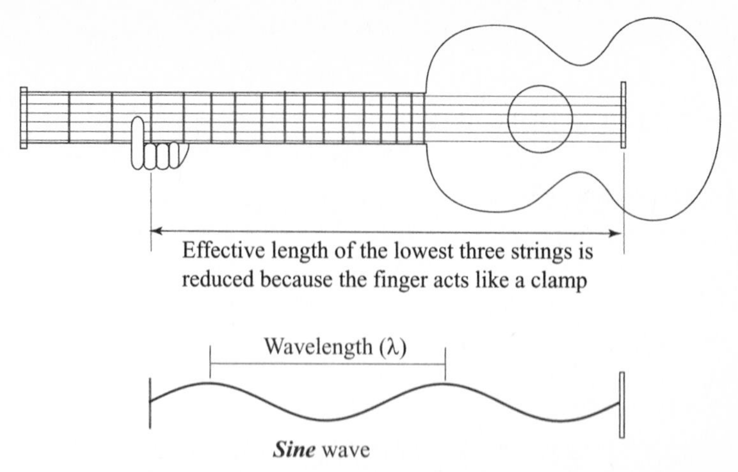

Figure 1.3 When a guitar string is plucked, it vibrates in the shape of a sine wave, characterized by its wavelength (denoted by λ), which is just the separation between adjacent crests of the wave. For each string, different notes are produced by varying its "boundary conditions" as its effective length is changed by clamping down with a finger. The effective length determines the allowed wavelengths as shown in the next figure.

duality[3]: Most quantum entities behave both like waves and like particles depending on how you look at it! Let us now see how waves and boundaries lead to quantization.

Consider the waves on a vibrating guitar string; high-speed snapshots show them to have the easily recognizable sinusoidal shape shown in Figure 1.3. We all know that we won't get far trying to make music with the strings hanging loose! The strings need to be clamped down at the ends and then tightened to the right tension to tune them. Beyond that, playing the guitar (or any string instrument for that matter) is all about varying the effective length of the strings as you pick on them; your fingers act just like temporary

[3] The *wave–particle duality* is one of the strangest features of quantum mechanics and shows that at very small scales all entities behave both as particles and as waves depending on the type of phenomenon in which they are involved. It's sort of like a Dr. Jekyll and Mr. Hyde situation—the same person but dual characters manifest in different environments, except that with waves and particles, there is no good or bad, they are both equally relevant.

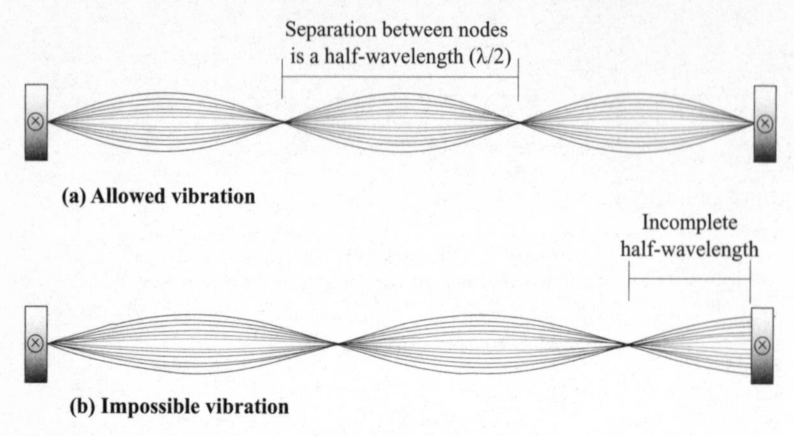

Figure 1.4 (a) The wave shown here is allowed by the boundary conditions of having the endpoints fixed. The equally spaced points (including the endpoints) that never move are the nodes. (b) The wave shown here is not allowed, since complete half-wavelengths do not fit between the endpoints, requiring motion of the string at the right endpoint, which is not possible since the string is attached there.

"clamps" where they press down along the neck. Everyone, from rock star wannabes to Eric Clapton does just that when pressing down on the frets along the neck of the guitar; bending the strings simply stretches them a bit more. The effective length of a string, illustrated in Figure 1.3, determines the musical note it plays by fixing the wavelength of its vibration. Let's see how.

If you were to look closely at any vibrating string, you would just see a blurry outline, as shown in Figure 1.4. But if the string is not too tight, you would also notice that certain points, spaced out at equal intervals along the string, never move at all! Those points are called *nodes*, and the interval between adjacent nodes is exactly one half-wavelength of the wave playing on the string.[4] Now here is the crucial point: Since the endpoints are fixed, and all the nodes

[4] The wavelength of a wave is the distance over which it repeats its shape (see Figure 1.3). The wavelength is a fixed characteristic of a wave; it is the same for any pair of crests of a particular wave. The nodes are a half-wavelength apart because the shape of the wave repeats exactly only at every other node. For more about waves, see Chapter 9.

in between are *equally spaced*, an allowed vibration of the string will need to fit *complete half-wavelengths* between the clamped endpoints. Incomplete half-wavelengths will not do, since they would require *impossible* motion at the *fixed* endpoints, as illustrated in Figure 1.4(b). Therefore, any string with its ends fixed can vibrate only at very specific wavelengths; the boundaries of the string determine the wavelengths allowed.

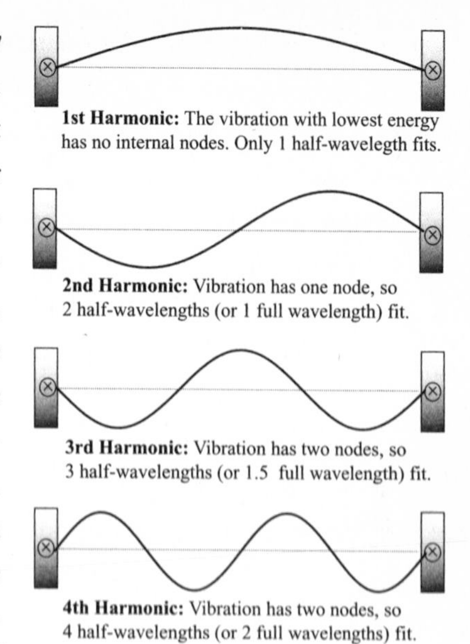

Figure 1.5 Snapshots of a vibrating string showing its lowest few stationary states (or *harmonics*). The nodes are points that never move during the specific vibration associated with each harmonic.

Even for fixed boundaries, there can still be many allowed wavelengths. But they are not arbitrary. Only such waves can play on a string that exactly fit a set of complete half-wavelengths between the endpoints, as shown in Figure 1.5. It's a bit of a Cinderella situation; only a select few fit the "glass slipper"! Those select few waves are the *stationary states* for the "boundary conditions" of a particular fixed length of string with its ends clamped down. The states are indeed stationary because, if there were no air resistance or friction, they would vibrate forever, and unlike "traveling" waves, such as ripples in the water, these waves do not go anywhere; they remain right there on the string.

In quantum jargon, the stationary states would be the *eigenstates* of the string. In musical parlance, the stationary states are called *harmonics,* the lowest one with no nodes being the *fundamental.* So there is a poetic connection between the quantized view of happiness to sound and music as well as to light! The higher the number of nodes, the higher the harmonic, and don't we associate higher harmonics with happier feelings?

The essential point here, that *boundary conditions determine the states,* directly translates to the fundamental rule of happiness in life: the happy or sad (or any other) state of our mind is defined by the boundaries, physical and mental, that frame our existence, meaning that our happiness and our general state of mind for the most part depend on our environment, the circumstances of our life, and the people around us. In any given situation in life, you can easily identify the key set of boundary conditions that constrain you: your connection to family and friends, your bank balance and line of credit, your job security, your home and neighborhood, your health, and such other relevant factors in your life, some major and others minor. With a fixed set of boundary conditions in your life, your options for happiness are restricted or "quantized." Change the boundary conditions, and your state will automatically change—as surely as musical notes change when the vibrating segment of a guitar string is stretched or shortened.

Boundary conditions of our lives, however, do not change often or easily, yet we all know that the state of our mind can and does vary day to day, hour to hour, without any significant changes in life situations. That is absolutely consistent with what happens in quantum mechanics; multiple harmonics are indeed possible within fixed boundary conditions as with the guitar strings in Figure 1.5. The catch is that higher harmonics require more energy to generate and are therefore harder to reach; for example, normal picking on a guitar usually generates only the lowest harmonics on the strings. This natural bias toward lower harmonics underscores the essential reality of life that we all live with: It is always harder to achieve a happier and contented state of mind than it is to slide into despondency and depression. Hey, nobody ever complained of too much happiness, while antidepressants have become a staple of the modern world—shrinks count upon nature's proclivity for the lower harmonics!

The message here is that there are two routes to happiness: If the boundary conditions in your life remain the same, then you can try to scale up to the higher harmonics of life powered by personal energy and self-motivation. Or you could instead change your boundary conditions—alter and improve the factors that define your life now—and thereby automatically upgrade your state of mind. The trouble with the second route is that many of the boundary conditions in life are often things that we can't control.

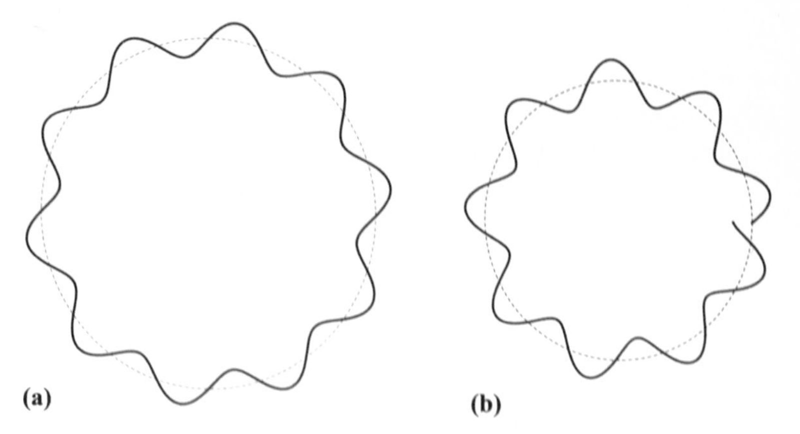

Figure 1.6 If the electron is thought of as a wave that is wrapped around in a circle, then it is easy to understand how boundary conditions lead to only a specific size of the orbits being possible. (a) The ends of the wave meet perfectly for this allowed orbit size. (b) The ends of the wave do not match for this slightly smaller orbit size, hence this particular size of the orbit would not be allowed.

Some say true happiness comes from within. Not quite, according to this quantum perspective. Rather, we can understand it to mean that if we can succeed in making ourselves happy without having to change our boundary conditions, then we are in charge, we need not depend on others and on external factors, and therefore we achieve a more lasting sort of happiness.

We can understand now why electron orbits are quantized. It is all about the wave–particle duality. Electrons behave like particles and like waves depending on what aspect of it we are looking at. Think of an electron as a standing wave, as on a guitar, only wrapped around in a circle, as shown in Figure 1.6. Instead of it having clamped boundaries where the boundaries cannot move, here instead we need the ends to match up smoothly on being wrapped around. This is a slightly different boundary condition (called *periodic* boundary condition), but the argument is exactly the same as for the guitar string. Instead of changing the wavelength of the waves, we just keep on adding more and more half-wavelengths. This means that to squeeze in exactly one more half-wavelength, we need to increase the radius by just the right amount to keep the endpoints matched up.

What is truly amazing is that light and sound—our favorite metaphors for describing the states of the mind—naturally blend into a quantum view of happiness. It is as if somehow we always were in conversation with the universe and understood its language without deciphering the words.

CHAPTER 2
THE HEISENBERG COMPROMISES

A cornerstone of quantum mechanics, the Heisenberg Uncertainty Principle, provides the ultimate answer to one of life's eternal conundrums: Why is it that we can never "have our cake and eat it too"?

Charting the best course through life is often about making the right compromises—a lesson we all learn sooner or later. Coming to terms with this fact marks the transition to maturity. We start life wanting it all and throwing tantrums until we get what we want—a brief phase that passes quickly enough as various trade-offs are forced upon us, innocuously at first by parents and siblings but then with increasing intensity as life goes on. And there is no getting around them, no matter who you are, or how lucky and powerful you feel. We all need to make compromises in life, and we need to make them often. Really now, what chance could we ever have, considering that trade-offs are built into the very fabric of the universe through one of the cornerstones of quantum mechanics: the *Heisenberg Uncertainty Principle*. It's a fundamental law with telling analogies that apply to all the inevitable sacrifices that we have to make in our lives. What the uncertainty principle says is deceptively simple and quite easy to understand, but the implications are so unsettling that even Einstein could never come to terms with it.

As a recognized fundamental law of nature, the uncertainty principle has been around for less than a century. However, the message of it has been ingrained in our traditions for ages. The English language, for example, has the old familiar adage, "You cannot have your cake and eat it too." Variations of that surely exist in all languages and cultures, so chances are you have heard or said something to that effect one time or another. If this particular phrasing gives you pause (after all, how do you eat a cake you never had?), well, that's because of its roots in old English. Turn the phrase around, "You can't eat your cake and have it too," and the message comes through—that you cannot continue to have the same cake tucked away in your kitchen after you have already eaten it. Well, cakes are nice, and we would like to eat them often and still have plenty to go around. But, of course, this idiom is not really about cakes. It is the distilled wisdom of the ages about the trade-offs in life—the undeniable fact that in most life situations, you cannot have things to your satisfaction in every way you want, like having a full cake in your possession after enjoying the taste of it.

Compromises never get any easier with practice, and the big ones always leave behind regrets and a sense that the cards are stacked against you. Whenever you feel that way, you might remind yourself of the uncertainty principle to put it all in perspective. Because, you see, it is not just your particular lot in life to make the hard choices; compromise is an inherent and ubiquitous feature of the universe we inhabit. But why are compromises so difficult? Mostly because we never completely know the consequences of our choices. And our potential for knowledge is exactly what the uncertainty principle is about.

Science, as we all know, is in the business of knowledge, and given all the progress we have seen in the last century, we may feel confidence that gaining complete knowledge of anything is just a matter of time and about crafting better tools. Well, "complete knowledge" is hard to define, so let us start with something diametrically less ambitious: "minimal knowledge." We can agree that at the very minimum, our knowledge of any object, animate or inanimate, big or small, requires knowing two things about it: (1) its position (where it is located) and (2) its velocity (how fast it is moving and in what direction; if it is not moving at all, its velocity would be zero).

We should be able to measure them both as precisely as we care to. In everyday life, we absolutely rely on that belief—our common sense demands it. For instance, a traffic cop expects to be able to use his radar gun to determine the velocity of a car and simultaneously pinpoint the location of the vehicle. If that were not the case, it would be a breeze fighting speeding violations in court. Imagine the cop, standing alongside of you in front of the magistrate at the county court, with typical serious demeanor making statements like, "Yes, Your Honor, the defendant was going 100 miles per hour, but I am not quite sure exactly where the violation occurred," or better still, "I observed the defendant speeding at the corner of Fifth Avenue and Forty-Second Street, but her speed could have been anything between 15 miles per hour and 100 miles per hour." The magistrate would have no choice but to dismiss your ticket and would be absolutely justified in recommending probation for the officer on account of insanity or inebriation on the job. Justified—because the magistrate shares the general expectation of all sensible people that the officer should be able to determine with certainty both the speed and the location of the violation.

Until the mid-1920s, that worldview, that at least in principle we can simultaneously gauge all the relevant parameters of an object with any precision, was accepted as a self-evident truth that was universally applicable. Then, that cozy view fell apart when the young German physicist Werner Heisenberg showed with his uncertainty principle that for a tiny particle like an atom or an electron, it is *impossible* to know simultaneously both its position and its velocity exactly. If you measure its position precisely, then you will lose some information about its velocity; vice versa, if you measured its velocity precisely, you will have to sacrifice complete knowledge of its position. Here's a simple way to understand why: If we make a very careful measurement to pinpoint the location of an electron, then the very process of that measurement will make its velocity uncertain. That is because to find its position, we will have to probe it somehow, like by shining a light on it as shown in Figure 2.1, and for something that small, even the light hitting it would give it a kick that would change its motion. As a result, its velocity now becomes uncertain. Of course, with larger entities, like you and me, light can't knock us around. But the principle holds; there is an eventual limit to how much we can know about any object in this universe. Our knowledge of even the tiniest particle in the universe is fundamentally limited by a quantum trade-off!

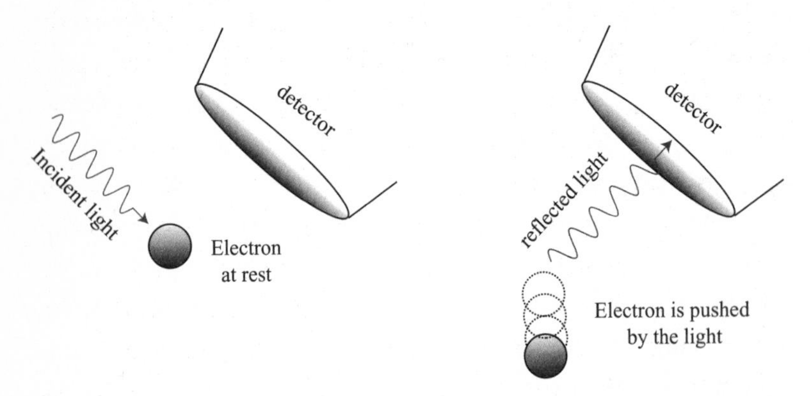

Figure 2.1 The process of determining the position of a subatomic particle, such as an electron, changes its velocity because even the energy of the light used to observe it gives it a kick.

But the strangest thing about the uncertainty principle is that all observable physical parameters come in pairs called *conjugate variables*, for which if one is measured very precisely, inevitably some precision is lost in the other. It is kind of like most married couples, both can never seem to be happy at the same time! Well, the name can't be a coincidence—after all, a married couple is a "conjugal" pair! Position and velocity happen to be just one such pair, but there are many others, like energy and time, for example. Interestingly enough, each member of a conjugate pair affects its partner, but not necessarily members of other pairs. So, for example, we can measure the position and the time both as precisely as we want to, likewise the momentum and the energy—no uncertainty trade-offs there. Now if that seems bizarre, just think about marital conjugal pairs; the happiness of a wife might correlate with her husband's but is independent of how the neighbors are feeling.

The uncertainty principle is a primary reason why the quantum view of the universe is so very different from the classical view. In the classical worldview defined by Sir Isaac Newton, you could know everything about any object in the universe completely and up to arbitrary precision—well at least in principle. This led to a deterministic view of the world, which strongly influenced the cultural and philosophical outlook in the centuries that followed Newton. The universe was viewed as a clockwork mechanism, so that if you knew the initial position and velocity of every particle in the universe and the forces acting among them, you could, in principle, predict how the drama of the universe would play out for all eternity.

The trouble with this deterministic view is that it did not leave any room for free will or matters of chance. If you bought into that picture, then you will have to believe that everything in the universe (including all our lives) were predetermined at the origin of the universe! So certainly some change of view was inevitable. Yet nobody could have expected the magnitude of the change that was to come! The quantum view *à la* the uncertainty principle really took us all the way to the other extreme: Now instead of the universe being completely deterministic, it turns out that, at the subatomic level, everything is governed by uncertainties and laws of chance. Forget about predetermining the fate of the universe, the uncertainty principle tells us that we cannot even precisely determine the fate of one single tiny electron!

In retrospect, the classical deterministic view of the world was an unrealistic one at many levels. For one, it simply does not agree with any of our life experiences. In a predetermined universe, chance can be eliminated, so you could, in principle, know it all. Well then, if you could know it all ahead of time, you could also arrange to have it all. Yeah, right! The reality is that in life we can never have it all, and our lives are indeed governed by chance, much more than we would like to admit. Life is chock-full of trade-offs, just like the uncertainty principle. And as with the uncertainty principle in nature, all the desirable and worthwhile things in life seem to come in conjugate pairs, with a built-in catch so that you can never get enough of both—because too much of one always means sacrificing on the other. In the language of quantum physics, we just say that *conjugate pairs do not commute.*

My favorite one is the "time versus money conjugate pair"; you never can seem to have all the money you want and the time to spend it simultaneously. Barring a few lucky individuals—those who came into large inheritances, like the idle aristocrats in the novels of P. G. Wodehouse, or those lucky few who made their pile young and retired early—the rest of us with high financial ambitions have to slog at it for most of our lives, leaving very little time to play with the money as we struggle to push up our net worth higher. On the other hand, if you have no financial ambitions, and you are completely content with living in a barrel—like Diogenes of ancient Greece—and with having nothing at all—like the naked fakirs in the Himalayas—you will likely have all the time in the world, but of course not much money to measure that time with. In the quantum jargon, time and money just do not commute!

Then, take the whole marriage thing, for instance. Not so long ago, the popular comedian, Chris Rock, a keen observer of human behavior, had this to say about marriage: You can either be "single and lonely" or be "married and bored." Without perhaps realizing it, he was touching a bit on the uncertainty principle, because I think that captures another one of life's great trade-offs. If you stay single, you can continue to play the field, you can hit the bars on the weekends, date several people in a single year at varying levels of intimacy, and move on when you are bored (assuming that you have the charm and the social skills to get someone to go out with you in the first place). At the end of every day, when you leave your workplace, you are free as a bird, you can do anything you want in your pad. If you happen to be a guy for whom a regular cleaning schedule is thing of mystery, there would be no wife to question your notions of cleanliness; if you planned for days to watch that game on Monday night, you would not be cajoled and manipulated into watching the mushy sitcom instead. And vice versa, if you are a woman who likes your nest to be organized and pretty, there is no one to mess up your place, no badly aimed urine all over your toilet, and if one night you felt like having a good cry watching *The Bridges of Madison County*, there is no danger of some insensitive lout switching channels just as Clint Eastwood is driving away from that last crossroad.

So, it is all great, you are completely free to enjoy yourself and live as you want to—that is, until you hit those weekends with no dates and cannot find any friends to hang out with; you are rained in and you do not even feel like watching television because it reminds you of all that is missing in your life; worse still, you are middle-aged now and getting dates is not as easy or as much fun as it used to be—and all your friends are married with kids and busy with their lives. Then you might start wishing for a family and kids yourself. You might have enjoyed the single life, but all along there was this risk of eventually ending up lonely and depressed. Not to mention that you probably have shortened your life by excessive partying, drinking, and smoking in the name of socializing.

Well, those are the trade-offs for the single life. There is plenty to say about the other side as well, the married life. I read somewhere that married guys are among the largest consumer demographics of the adult industry in America. So I am assuming all is not so quiet on that domestic front. For sure, married people envy their single friends often, and perhaps that is why they always try so hard at matchmaking; you know that if you are single after a certain age, and most of your friends are married, they will harass you about getting married, and they will keep on trying to set you up with someone—usually with people you do not want to meet. I think married people simply cannot bear to see all the fun their single friends are having that they themselves are missing out on, while they are doing household chores amidst whiny kids and intolerable spouses. So they try to ruin that fun for their single friends, by getting them married, as well. But hey, if you can survive all that and make it through the tough times, you will have very little time to get depressed and lonely, partly because you will never be alone for any length of time to indulge in either. And in your middle age, when the kids are off to college and doing well, you will be so proud and happy, and the spouse of many years has now become someone you cannot imagine living without. So when your still-single friends, the ones you used to envy a decade ago, stop by, they will linger over the photographs of your kids and wonder what went wrong in their lives.

While on the topic of family life, one can also apply the uncertainty principle to having kids. Sure, they can make you smile and make you proud and might rally around you (if you are lucky) in your old age, but then you have to be prepared to go through those sleepless nights and the trips to the doctor, the PTA meetings, soccer practice, piano lessons, and pretty much sacrifice many of the best years of your life to take care of their needs while they grow up.

Other trade-offs in life have a closer semblance to another set of conjugate variables in quantum physics. The energy–time uncertainty principle is just as fundamental as the position–velocity one we talked about. And in quite the same way, the more precisely we specify the total energy in a system, the less precisely can we define the time scale associated with it. This is somewhat trickier than the other sets of uncertainty-bound conjugate pairs in physics, because unlike position or velocity or even energy, time is not a characteristic of an object, be it an electron or a human being. Time, like space, is more like a background against which all other things play out. One way to understand the energy–time uncertainty principle is as follows: All fundamental particles have some intrinsic energy as a part of their quantum identity, just as your height, eye color, and other characteristics identify you. And the more precisely you want to know the energy of a particle, the more time you need to do the measurement. Therein lies the trade-off: The more precise your knowledge of the energy, the less precisely can you pinpoint the time when you actually have that knowledge, and, vice versa, the less time you spend in measuring that energy, the more precisely you can specify the exact moment of the energy measurement, but then you have to sacrifice the accuracy of the energy value you have.

As you will see often in this book, the human implications and the life-parallels of many of nature's laws have already been distilled (albeit unwittingly) into some of our traditional aphorisms. A real-life analog of the energy–time uncertainty principle is the well-known figure of speech, "jack of all trades, master of none," because effortless virtuosity requires time and dedication. To master anything—to minimize the uncertainty in your knowledge of a subject—you need to dedicate an immense amount of time; whereas if you are fickle and spend only a short time on any particular interest, you will be just another jack at it, never a master. Beethoven was the genius that he was because he had the dedication to perfect his art, and he sacrificed much in his life along the way. Einstein worked on general relativity for ten grueling years. Thomas Alva Edison was a workaholic who only slept a few hours a night, and his long and persistent efforts changed our world. A famous musician once commented wryly about his skills that, "It is not easy to make it look easy." That is so very true. There is no uncertainty in virtuoso performers playing their instruments; they are sure of how to play and what to play—it all flows. But behind it all, lies years and years of hard work and sacrifice and obsessive drive to get there.

Well, I could rant forever about all the trade-offs in life, and I am sure you can come up with many more of your own, but here are just a few of my favorite ones to convince you that the spirit of the uncertainty principle indeed defines our lives:

- The bigger and more lavish the wedding, the shorter the marriage lasts; think of celebrity weddings!
- The more magnificent the house, the less time people seem to spend in it. Many large mansions are owned by rich folks who own multiple homes and only spend a few weeks or months at each. On the other hand, the small suburban houses with kids stacked on bunk beds seem to be always full of people and life.
- The more spectacular the kitchen, the less cooking is usually done in it. The best cooking always seems to come from your grandma's humble, old-fashioned kitchen; whereas in that

beautiful thousand-handle state-of-the-art kitchen, the only handle that is probably used is the one for the microwave.

- Women complain about guys being either interesting and attractive, but unfaithful and deceitful, or very dependable and reliable, but boring and unattractive. Guys have similar complaints as well; women they can easily get are never the ones they are attracted to, and the ones that are attractive are always out of reach.
- You like the peace and security of the country—great quality of life and great for raising kids—but then you risk becoming bored out of your mind often and missing out on all the culture and excitement of the big cities. A place in the scenic boonies is great until you have an emergency that requires urgent care. Move to the city, and you could get all the excitement and activities you could ever want, but then you need to deal with traffic, crime, and a whole bunch of other everyday hassles.
- You get drunk and feel at the top of the world for a few hours, and then you wake up with the worst hangover the next morning—and perhaps next to some loathsome stranger!
- You are the cool kid in your high-school class, spend all your time partying and chasing girls, could not care less about your studies, and think life is so good that you might even drop out of school—but then you risk ending up like Al Bundy in *Married with Children*. The nerds and geeks who seem to have no life, no dates, and are bullied daily at school are the ones who might end up as the Bill Gates or the Steve Jobs of the future.
- You have a lot of stuff and you always have to worry about losing it, or you don't have a care in the world—and you probably have nothing to lose, either.
- Throughout history, people who settled down in villages and towns, nations and countries, could count on a better quality of

> life in general. But when the nomadic tribes, the so-called barbarians, invaded, the barbarians had a huge tactical advantage, and they usually won with devastating success and then disappeared with all the loot. The cities and nations could, and did, send out huge armies after them, but could never find them! This was true when Rome fell, when Genghis Khan ravaged the known world, and it is true now, as the civilized nations try to battle terrorists and combat guerilla warfare.

But at the end, a bit of uncertainty in life is perhaps good for us. That is what makes life exciting and interesting. It is also the source of hope. You might not want to know everything about everything—the reality can be quite painful sometimes—like what the future really holds or what people really think about you. Besides, much of our plans and dreams of life are built upon the premise of uncertainty and the absence of complete knowledge. Romance is all about uncertainty; without it, there would have been no *Casablanca* or *Doctor Zhivago* or the powerful final sequence of *The Third Man*. In an absolutely certain world, all the romance you will have will be children's fairy tales. Likewise, there would have been very little drama and suspense; Hitchcock might as well have been making cartoons. And not just in movies, even in real life, doesn't much of the excitement and romance in our life ultimately arise from innate uncertainties that surround us?

If uncertainties of life ever get you down, it might help to remember that it is not just your life; uncertainties define the universe at the most fundamental scale. Even in the world of physics that purports to be so exact, in the end we're really working with probabilities. It is all a game of chance from the smallest scale to the scale of our everyday lives, and we just have to weigh the probabilities, as there is always a trade-off; it's just that in physics, we have formulae for those trade-offs. In real life, we have to make ours up as we go along.

CHAPTER 3
ENTROPY IS MESSING WITH YOUR LIFE

The second law of thermodynamics is the most fundamental law of the universe, and in it we will find the real cause of all the disorder and disarray in our lives.

A messy office, the challenges of parallel parking, the universal need for freedom, exorbitant winter heating bills, metastatic cancer, and the arrow of time—what do all of these have in common? A seemingly disparate and random mix, they are all related to the most fundamental physical law in the universe: the *second law of thermodynamics*. It affects everybody and everything that exists, yet it has a deceptively simple statement: *the net amount of disorder in the universe always increases.*

You just might have been blaming all the wrong things for the mess in your life. The true culprit is the second law of thermodynamics. In the entire hierarchy of laws, this is the ultimate one—the one law that nothing and no one in the universe can violate. It is nature's equivalent of the presidential veto, with the final say on what is allowed and what is not. If you ever had visions of going back in time and mending your mistakes, or had wistful thoughts of perpetual motion, or set yourself goals of absolute efficiency and everlasting order—you can forget about all that, because you won't get past the second law of thermodynamics. It is so sacred that all our current understanding of just about everything would come crumbling down if any exception to this law is ever found. And we are all in its grip. If ever life felt like an uphill battle to keep things under control and hold chaos and disorder at bay, you are now face-to-face with the real reason why. If you get to know the law, you might be able to loosen that grip a bit by pushing the limits—or at the very least, learn to be somewhat more forgiving of yourself and others when perfection is elusive and out of reach.

This law is messing with your life—literally. Without it, life would have been a lot easier. In fact, life would have been downright magical! Much of what is passed off as magic in books and movies of fantasy is actually based upon violating the second law. It is because of this law you never see a broken glass put itself back together again. It is because of this law you can never expect to wake up one fine morning to find that your room has spontaneously organized itself. Say you went to sleep with your clothes on the floor, soda bottles under the bed, magazines and half-read books randomly strewn around the room; you locked your door and went to bed, and no one came by during the night. Then, you wake up the next morning and look around to find that your clothes have folded themselves neatly in the closet, there is no trace of any empty bottles, all the magazines and books have very considerately closed themselves shut and arranged themselves on the shelves . . .

Now wouldn't that be nice! But it never happens, does it? It never happens because the second law of thermodynamics demands that the overall disorder in the universe always increases. And to imagine that all this time, you might have felt with exasperation that disorder conspires to specifically target your house and your workspace! Disorder has no favorites—it will grow and increase anywhere and everywhere it can—and because it is so ubiquitous and naturally rampant everywhere, physicists even gave it a special name of its own: we call the level of disorder in a system *entropy*. We can call it by any fancy name we want, but there is no mystery really about disorder and entropy; disorder in physics means exactly what we understand by the word in everyday life. Just as with my little sketch above about the messy bedroom, we know only too well from experience that in general if we don't make any effort, any bunch of things would eventually become more disorderly and messy. Now, you can officially stop blaming yourself and the people around you for all the mess, because it really has got nothing to do with human behavior *per se*. It just happens to be a very basic law of the universe, the most basic, in fact. What's more, if people have had the audacity to accuse you of being messy and keep on harassing you to clean up after yourself, well, next time you can tell them that they can pack up and move to another universe to live in; as for you, you are only following the local rules.

Even a hundred years ago, the primal significance of the second law of thermodynamics was already well recognized, as expressed in this wry statement from the renowned British scientist, Sir Arthur Eddington, the man who made Einstein a household name:

> *The law that entropy always increases, holds, I think, the supreme position among the laws of Nature. If someone points out to you that your pet theory of the universe is in disagreement with Maxwell's equations—then so much the worse for Maxwell's equations. If it is found to be contradicted by observation—well, these experimentalists do bungle things sometimes. But if your theory is found to be against the second law of thermodynamics I can give you no hope; there is nothing for it but to collapse in deepest*

> *humiliation.*—Sir Arthur Stanley Eddington, *The Nature of the Physical World* (1927).

Well, the gist of it is that any scientific theory that violates the second law is utterly doomed, a point Eddington makes by saying that it is even more sacred than Maxwell's equations, which in themselves are pretty inviolate and sacrosanct in the canons of physics because Maxwell's[5] four equations precisely describe all of electromagnetism and the nature of light everywhere in the universe from the stars and the galaxies down to the inside of an atom. In fact, Maxwell's equations are so fundamental that when pitted against the venerable and well-tested theory of mechanics laid out by Sir Isaac Newton, it was Newtonian mechanics that had to be altered as Albert Einstein ushered in the *theory of special relativity*. Yet almost any physicist will readily agree that Sir Eddington is absolutely justified in saying that the second law of thermodynamics is even more fundamental.

The origin of the second law is actually quite easy to understand in common sense terms. You see, it is all about available possibilities and probabilities. Ask any teenager trying to pass the driving test, "which part of the test do you dread the most?" and the answer would almost inevitably be, "parallel parking." Parallel parking is never trivial, and even some of us with aging licenses might still be challenged when parking a car in the streets of New York City or the narrow alleys of old European cities, where cars are literally parked bumper to bumper on the street side. But no one ever finds it very challenging to pull out of parallel parking. Why not? Why is it so much more difficult to parallel park than to pull out of parallel parking? In a way,

[5] James Clerk Maxwell was a Scottish physicist from the mid-nineteenth century who demonstrated that electricity and magnetism, previously thought of as two distinct phenomena, are actually intimately related and are describable by four compact equations that bear his name. These equations essentially unified electricity and magnetism into a single description that is now known as *electromagnetism*.

this has also got to do with the basic idea of entropy. There are many, many more ways you can be outside of the parking spot than you can be in the parking spot, or in the language of physics, there are more possible "states" outside than there are inside. That is exactly at the heart of the second law: There are zillions of ways to be disorderly, but sometimes only one way to be orderly. Thus, any system naturally has a much higher probability to be in one of the many disorderly states than in the orderly one you might prefer. For example, if you fill up a box with black and white marbles as shown in Figure 3.1, you are

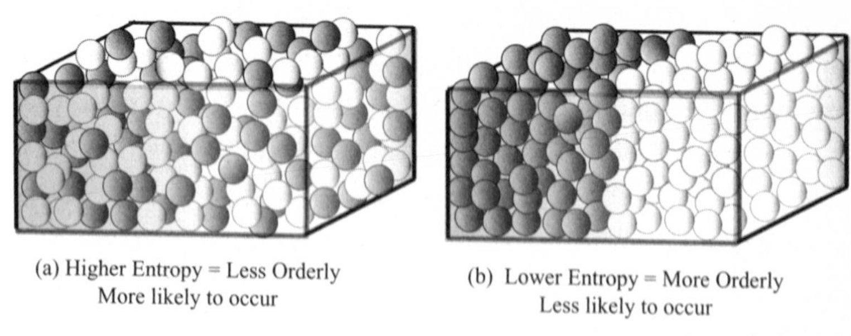

(a) Higher Entropy = Less Orderly
More likely to occur

(b) Lower Entropy = More Orderly
Less likely to occur

Figure 3.1 The more disordered outcome (a) where the balls are all mixed together has higher entropy and is much more likely to occur than the less disordered one (b) where all the dark balls are on one side and light balls are on the other.

unlikely to get all the black ones on one side and the white ones on the other, because there are many more ways for them to become mixed up together. And, therefore, with time, disorder prevails and entropy inevitably increases.

Actually, you would be surprised at how much of what you always thought as quintessentially human traits is dictated by this law. Take our universal and deep desire for freedom and liberty, for instance. We have heard, "Give me Liberty, or give me Death." But why are we so obsessed with freedom? Why are we willing to court the greatest of privations, and even death, for a taste of freedom? There have been infinitely many poetic and psycho-sociological answers through the course of history. Now, here's a scientific answer: We all like freedom because freedom allows us the possibility to maximize our personal entropy. Being in a maximum security prison cell is a human condition in a state of very low entropy because the location is pretty much fixed and determined; it cannot change in a disorderly and random way over time. Nobody likes to be in such a state. You would probably take your chances with getting stalked by the mafia and dodging bullets from hired assassins out in the open world rather than being stuck in the relative safety of a prison cell for the rest of your life. At the other extreme, you could be a billionaire playboy with an unquestioning indulgent parent taking care of the family business, and you could have a private jet at your disposal to travel anywhere at any time; that is a condition that can lead to very high entropy in terms of your coordinates over time, and who would *not* want that? The need for freedom could be understood as a manifestation of the second law of thermodynamics. That need is as fundamental for us as the second law is for the universe in general.

However, we are not completely at the mercy of the second law. Although the overall entropy in the universe has to increase, one can reduce entropy in any little corner of it, but always at the expense of raising it *even more* elsewhere. And reducing entropy in any small corner always comes at a price, and that price is *energy.* Expenditure of energy is necessary to reduce entropy of any particular system. Some familiar examples will illustrate that.

Even if you have never heard of entropy, you have probably measured the level of entropy many times in your life without realizing it. Yes, you have indirectly gauged entropy every time you measured temperature, because, you see, temperature is a measure of entropy. Consider the air around you. The air is made of tiny molecules of oxygen and nitrogen gases mainly, and these molecules are constantly moving around at random. If the air gets hotter, these molecules move around faster in an even more disorderly way. The same is true for anything that you can take a temperature of: the hotter it is, the more intrinsic entropy it has. So it should now be easy to see that there is a connection between energy and entropy. Your heating bills in the winter or the electricity bills to operate your kitchen range, microwave, and iron should make it painfully clear that you have to spend energy to increase temperature and therefore entropy.

But what is not so obvious is that energy is also required to *reduce* the entropy of an object. So entropy gets you both coming and going! Another household appliance will illustrate this: the refrigerator. A refrigerator cools down its interior and anything it contains, reducing entropy therein. But if you go to the back of the refrigerator, you will find that it is pretty hot because the refrigerator is pumping the heat from its inside to its outside—into the kitchen it is in. So entropy might have gone down inside the refrigerator, but that happens at the cost of increasing it even more outside. If you could add up the entropy inside and outside, you will find that there is a net increase in the entropy as dictated by the second law of thermodynamics. Your refrigerator also needs energy to operate—a lot of energy in fact. It obviously does not work unless you plug it into a power outlet. So the point is, even to reduce entropy of a particular system (like the inside of the refrigerator), you need to spend energy. That reminds me that once I heard some DJ on a radio jokingly advising his listeners one hot summer day to open all their refrigerator doors to keep cool. I hope no one tried this fool's advice at home, because it would never work, and, in fact, it would make your room even hotter, because the cooling effect is more than compensated by the heat from the back of the refrigerator. For the same reason, air cooling systems in houses always have a part outside the house, because to cool the house, to reduce the entropy inside, they need to dump the heat and excess entropy outside.

That we need energy to reduce entropy is what requires so much work from us just to get by in life and survive. A big fraction of our life and energy is spent in simply reducing entropy one way or the other. Life is a constant battle against the second law. Examples abound: to clean up your room, you have to do some work using up that valuable energy you scraped together from your breakfast. Parents need to use serious mental, vocal, and sometimes even muscular energy to restrict their kids to a preferred state of lower entropy, like being safely tucked into bed and asleep by 8 p.m. and remaining there till the morning without any rebellious desire to increase entropy in the wee hours. Teachers take over from the parents in kindergarten, and in turn expend mental, vocal, and muscular energy to prevent the kids from escaping into higher entropy states outside the classroom and the school boundary. The very existence of life in any form is a defiance of the second law of thermodynamics. Living organisms are organized and orderly entities, islands of order in a universe of disorder, and the food consumed is the energy input for maintaining those little blobs of reduced entropy. In recent years, there have been some intriguing suggestions that the origin of life itself might have been triggered as a means of accelerating the rate of entropy increase because to create and maintain the delicate order within itself, a living organism has to dump entropy into the environment at a significantly higher rate than possible by most physical mechanisms in systems of similar size. Evolution is then simply a way to create new species that increase entropy more rapidly, so from such a perspective, we humans seem to have exceeded nature's expectations!

The second law actually has some great news for all calorie counters out there! Diet peddlers love to scare us with how many calories there are in every donut or burger you reach out for, giving the guilt-laden impression that all those calories are going straight to our bellies and our bottoms as fatty padding. In reality, the body needs a lot of calories just to overcome the second law, just to keep us alive. So if you are calorie conscious, the situation is a lot better than you might have thought—you do not need to run a mile to burn every hundred calories you eat; your body naturally burns calories even if you lay in bed all day, just to keep you living and breathing, just to overcome the second law.

The second law applies to all aspects of life. Take finance, for example. One of the common bits of cliché wisdoms in the world of finance is that "there is no free lunch," which implies that you cannot get something for nothing. There is a price to pay for everything you gain, although it may not always be obvious. It savors strongly of the crisis faced by many inventions of perpetual motion machines, which seem to work at first but somehow always violate the second law of thermodynamics, which also implies that there is no perfectly efficient process—increasing disorder makes sure of that.

Aging and gradual decay of living things is essentially the eventual capitulation to the second law of thermodynamics. If staying alive did not require energy, and if disorder did not rule, it would have been a lot easier to approach immortality. Different creatures would not need to devour and consume each other to stay alive, microbes would not invade and weaken our bodies for their survival, and cancers would have had a harder time spreading rampant disorder through the delicate orders of our bodies.

Time itself is defined by entropy and the second law. If you ever dreamed of time machines, like I have, there is good news and bad. We can time travel into the future, as we will see in Chapter 16. But we can never go back in time. It has nothing to do with paradoxes, as in the parents not hooking up *à la Back to the Future*. In fact, about all[6] the equations in classical and quantum mechanics and electrodynamics are invariant with respect to the direction of time; they work just as well with time going backward. As far as they are concerned, we should be able to go back in time as well forward. It is the second law of thermodynamics that sets the "arrow of time." The universe as we know it can only evolve toward the direction of increasing entropy, and that is toward the future. Going backward in time would imply reducing the overall entropy of the universe, violating the second law of thermodynamics. That is forbidden—there is no going back in time, I am sorry to say, not in this universe. Perhaps we could jump to some parallel universe some day in the distant future where the second law runs backward, but that is all speculation and science fiction really. Don't count on it!

Really, technology has made so much happen already that would have been things of magic not so long ago, that today a general belief pervades the worldview of most educated people that given enough time, anything is possible. Therefore, it is worth reminding ourselves that certain things will never be possible, because anything that violates the second law of thermodynamics, anything that reduces the net disorder in the universe, is unlikely ever to happen. How to reduce the net entropy in the universe will always remain "The Last Question"—the title of my favorite Isaac Asimov story (and apparently his own favorite self-penned one, as well), which makes that very point beautifully by following eons of human evolution and technology always faced with that one unanswerable question that can be phrased, "Can the workings of the second

[6] There is one exception. One of the fundamental forces of nature, the *weak nuclear interaction*, breaks the time-reversal symmetry. But while this is tied to specific dynamical interactions inside the atom (giving different outcomes going backwards or forwards in time), the setting of the direction of time by the second law of thermodynamics has its roots in the origin of the universe itself.

law of thermodynamics be reversed?" It ends with a surprising final answer that makes it well worth the read!

But just because something is impossible has never kept us from dreaming. We will forever dream of breaking the shackles of increasing entropy, we will continue to write books and make movies and television shows about magic where people go back and forth in time, where they effortlessly create something tangible and orderly out of disorderly intangibles, where people live forever, and where it is routine for impossibly shattered things to mend themselves spontaneously, and cleaning and organizing are just a wish away. Well, in a way, it is comforting to know that we can always dream and spin stories of magic and fantasy, because there will always be some magic that science and technology will never reach. The second law of thermodynamics will always remain magic's last stand.

CHAPTER 4
THE LAZINESS CLAUSE

The roots of our inherent laziness and endemic impatience go much deeper than we tend to think. They are no mere human shortcomings; rather, the blame can be placed squarely on the laws of physics, most of which result from insisting upon least action and least time to accomplish anything.

When you hear *the principle of least action*, what pops into your mind? Perhaps a lazy relative or a lethargic college roommate who has embraced it as the guiding principle in life. There are certainly plenty of folks out there for whom getting by with the least amount of action and physical activity can be akin to religious dogma. Just the other day, while driving me back to his place in New York City, a friend announced with visible pride that he considers it a daily personal challenge to find the closest available parking space near his home. He was a man of his words, so we drove around the neighborhood blocks for more than ten minutes until he found a space to his liking. I did not have the heart to tell him that it would have taken us only two minutes to walk the distance to his house from the first spot we had found.

In matters of least action, we are all complicit. Even the most active among us look forward to lazy afternoons with nothing to do, when we can dig up some snacks, pour ourselves a drink, and sink into a couch with a favorite movie or a book. Taking pleasure in such states of low action is natural enough. But every so often, we just go overboard in our pursuit of minimal action, don't we? This occurs when we carry the essentials from the kitchen to the couch, all in one trip, balancing items like a circus act, risking spillage and broken plates just to avoid making a few extra trips to the kitchen.

Such all-too-familiar antics underscore a fundamental truth about human behavior: We usually seek out ways to invest the minimum amount of action and active work to get anything done, like walking a lesser distance from the car or making fewer trips to the kitchen. This is commonly attributed to human instincts and subtle consequences of conscious or subconscious decisions of the mind. But, in truth, human instincts or intelligence have very little to do with it; we are simply submitting to a fundamental law of physics, because, it so happens, the universe demands minimal action from pretty much everything that exists. Yes, that's right: there is a *laziness clause* built right into the fabric of the universe! Do you realize what this means? This means that if your daily quests for reduced action and effort have ever drawn accusations of laziness from judgmental friends and relatives, now you have the perfect defense: "Stay off my back—I am just an innocent victim of the laws of physics." If that's met with raised eyebrows and skepticism, just tell the doubters to read this chapter, because the next few pages here will make your defense watertight.

The *principle of least action* is truly among the most fundamental principles by which all creation works, and we humans just happen to be pawns to that rule. At the end, it is a good thing really, and we should be immensely grateful for it, because one could argue that our entire technological civilization is driven by this laziness clause. After all, most of our inventions, starting with the wheel and assorted stone tools, were motivated by our inherent desire to reduce the amount of work we do—to cut down on all the action we have to invest in to survive, to be comfortable, and to get places. So, to all who are quick to decry laziness, you might also add that they should learn to appreciate your inventive approaches for minimizing your actions and reducing your workload, because these just might be the seeds of genius—the kind that led to the steam engine and the motor car. A great way to show that appreciation would be by cleaning up all possible mess caused by your effort-saving antics. Of course, with such suggestions, you run the risk of being chased out of house and home, like many other inventors through history. But you would have truth on your side: The principle of least action is so fundamental that we can deduce *all* of classical physics and almost all of quantum physics from some condition that minimizes *action.*[7]

In case you are wondering what minimization means, it means exactly what the name suggests: We try to find the conditions that will lead to the lowest or minimum value of the quantity in which we are interested. For example, if you were trying to minimize your monthly expenses, you would look for all the things you could do to cut down costs to a minimum, like surviving on an exclusive diet of cup-a-noodles, riding your bicycle to work even with a foot of snow on the ground, and scrounging all your furniture from neighborhood sidewalks—now that is beginning to sound awfully like my life as a graduate student!

But what is the big deal about *action minimization* anyway? Why is it so special? Because action minimization holds the key to the simplest question of all and also the hardest one to answer: "Why?" You know that kids can drive us crazy sometimes by simply repeating "Why?" because we quickly run out of answers. In science, as in life, very often the simplest questions are the hardest to answer, and nothing is simpler

[7] Strictly speaking, the action is "optimum," allowing certain other possibilities than an absolute minimum, but that is a technical detail we will leave for physics textbooks.

to ask and harder to answer than the simple "Why?" In comparison, it is relatively much easier to figure out "How?" Modern science has become quite good at doing that; we observe, we conduct experiments, and from that we can figure out how things play out in nature. For example, we all know *how* a ball behaves if it is thrown up: It slows down and eventually turns around and falls back to the ground. There are equations that will precisely predict how the ball moves from the time it leaves the hand till it hits the ground. But if we were to ask why things always fall downward and not upward or sideways, that is a much harder question to answer. Following Sir Isaac Newton, we could say, "Because an invisible force called gravity pulls it down." But, like a persistently curious kid, we could ask, "Why does gravity pull things down, rather than push them away?" That is not so easy to answer.

Thus, science has established a systematic approach, based on careful observation and experiment, to figure out *how* nature works. But in the end, we are still left with the more profound question: "Why are the laws of nature the way they are?" Well, there might be no final answer to that, but couldn't we at least find some intuitive underlying principles that determine those laws? Amazingly enough, in most cases, the answer is, "Yes!" You see, almost all the correct laws of nature *obey certain minimization rules*, meaning that the true natural laws will be the ones that give the lowest possible value of some physical entity; all the unrealistic and impossible natural laws will lead to a higher value. There are minimization rules for several entities, including energy, time, action, path, and curvature. But they are all kind of tied together and connected really, and at the *heart of it all is this principle of "action" minimization.*

So what is this thing called the "action"? The physics definition of action is related to, but not exactly the same as, it's meaning in everyday usage—and it is rather technical. Yet it is easy to get an intuitive appreciation of it, because most of us are actively trying to minimize something very similar to the physical "action" in an important arena of life—and it is not what you might think—it is in how we handle our finances. You will see that the essence of sound financial practice emulates an "action minimization" principle taken straight from nature's rulebook on how everything in the universe should move and evolve in time. The "financial action" principle will ease our way toward understanding nature's version.

If you are a financially responsible person, and, like most people, you have to earn what you spend, you would certainly like to keep your account balance comfortably high and expenses relatively low. That is just the basic financial rule: live within your means. Here's another way of saying that: it is smart financial practice to *maximize* the difference, *"Bank balance" minus "Expenses."* But of course neither your expenses nor your balance remains constant; some days you spend more, some days less, some days you withdraw, and some days you deposit—both expenses and the balance will fluctuate day to day—so we just cannot go by daily values. Therefore, to truly assess your financial situation, we need to look at the *average* over a full financial cycle, like a month or a year. It is your average behavior that counts, meaning that good financial standing requires *maximizing* your *Average ["Bank balance" minus "Expenses"].*

Let us consider a concrete example. Say you have $3,000 in your checking account on the first of the month, and all your expenses are on your credit card, on which you owe $1,500, due on the twenty-sixth of the month. Since banks pay you interest (however miniscule) on your bank balance, it makes sense to pay out on the very last day. As shown in Figure 4.1, that would be your financial path through the month that *maximizes* your *Average ["Bank balance" minus "Expenses"].* Any other path, (I show three of infinite possible ones) where you pay all or part of the credit card bill earlier than the due

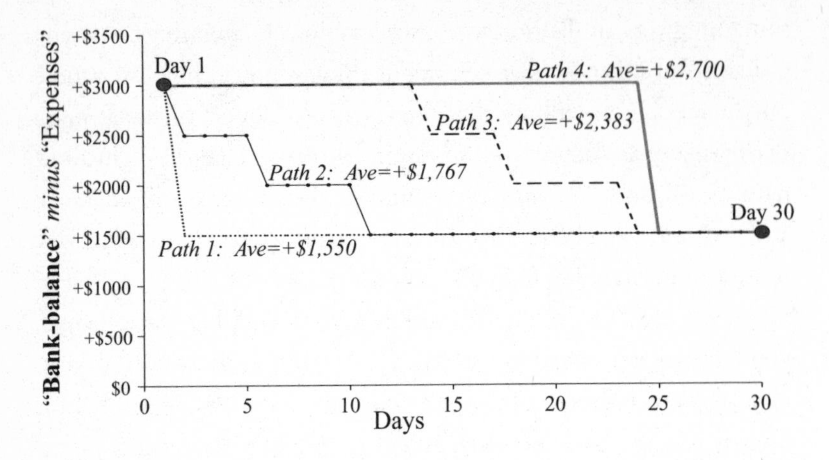

Figure 4.1 The best path is Path 4, where the entire balance of $1,500 is paid on the due date Day 26, so it gives the maximum value for the Average ["Bank-balance" minus "Expenses"] = $2,700. All other paths lead to lower averages as seen: along Path 1, all is paid on Day 2; on Path 2, payments of $500 are made on Days 2, 6, and 11; and on Path 3, equal payments are made on Days 14, 18, and 24.

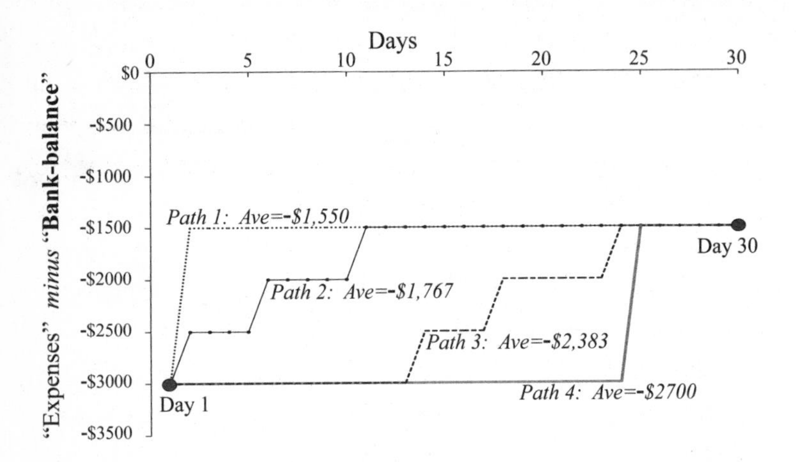

Figure 4.2 This is the same as in Figure 4.1, except everything is upside down, so all the values are negative instead of positive, because we are now subtracting "Bank balance" from "Expenses" instead of the other way around as in the previous plot. The paths are also the same as before, but upside down. Path 4 remains the best path, but now (being negative) it has the minimum average = –$2,700.

date, would give a lower value of the *Average ["Bank balance" minus "Expenses"]*, meaning lower interest income and lesser liquidity.

Now suppose the guys at the mortgage company who are investigating your financial habits have an algorithm that looks at the same thing *but turned around backward: Average ["Expenses" minus "Bank balance"]*

Since this switches what is being subtracted from what, this also flips all the numbers around so that all the "**+**" become "–" as you can see in Figure 4.2. But if you compare the two illustrations, you will see that all the different paths remain the same, except that they are all upside down in the second figure, and therefore the best financial path now corresponds to the *lowest* (or most negative) value of the *Average ["Expenses" minus "Bank balance"]*. It is important to realize that this is simply a different way of saying the same thing (just as "increasing your wealth" is the same as "decreasing your poverty"); therefore, it does not alter the choice of the best financial path, except that it now corresponds to *minimizing* what we will now give this specific name: ***Financial Action*** = *Average ["Expenses" minus "Bank balance"]*.

Thus, the optimal financial trajectory for most of us involves minimizing this *financial action*. But with credit cards and loans, we have found ways to bypass this rule, which is as fundamental in finance as its analog is in nature (as we see next), and if we are too reckless in bypassing this rule, in the long run it can get us into serious trouble—such as foreclosures or bankruptcies!

Now to understand how this "action minimization" principle works out in nature, we need to identify what is nature's equivalent of money. That's simple—it's energy! Just as good financial practice suggests that we should be thrifty with our money, nature is thrifty in its usage of energy, and it so happens *that this energy-thriftiness of nature determines all the dynamical laws of the universe* as regards to how things move and change in time.

Analogous to money saved in your bank balance, stored energy in nature is called *potential energy*—so called because of its potential to do work when released. Potential energy can take many forms: electricity stored in a battery has potential energy, which can power appliances; water piled up behind a dam has potential energy, because when released, it can generate power; the sun and the stars function by releasing the potential energy stored in the nuclei of atoms; a compressed spring has potential energy because it can be made to pull or push something when released.

Energy that is released or *expended* to do work and to generate motion is called *kinetic energy*—fittingly so, since the word "kinetic" relates to motion. Thus, kinetic energy is the analog of "expenses" in nature's book. Just like we can have a bank balance and expenses at the same time, objects can have both kinds of energy. For example, a spring that is oscillating back and forth is in motion and also undergoes compression and expansion, or a ball rolling down a hill is in motion (kinetic energy), but it is still not at the bottom

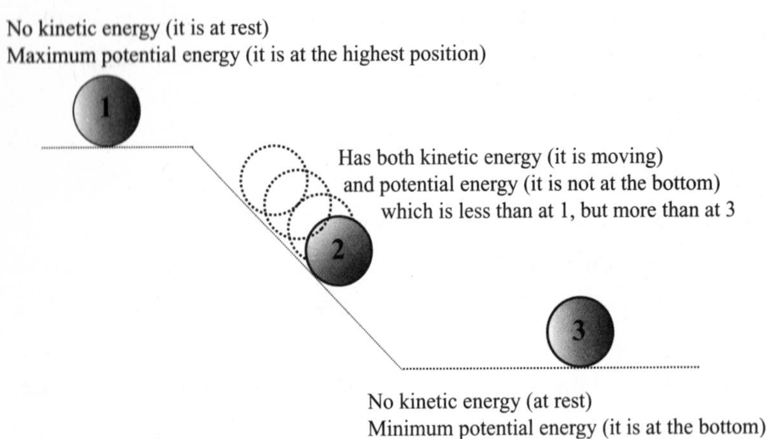

Figure 4.3 In position 1, the ball has the maximum potential energy due to its height, and it expends that into kinetic energy as it rolls down the slope.

and therefore has the potential energy to move some more—this is shown in Figure 4.3.

Now, we have all the ingredients in place to define the "action" as understood in physics. In exact analogy with the "financial action," in physics the action associated with any object in motion is defined to be ***Action*** = *Average ["Kinetic energy" minus "Potential energy"].*

The average is taken over the duration of motion and is necessary for the same reason that we needed to average the financial action over the number of days in a month: Since both kinetic and potential energy of an object can change over time, it is no good trying to minimize their difference at one time only to let it get very high at another time.

Nature's *action minimization* principle works exactly the same way as its financial equivalent. Consider a cannonball shot by a cannon on the edge of a cliff as shown in Figure 4.4. If all conditions remain exactly the same, we can be certain that every time a ball is fired off, it will follow the same trajectory (the solid dark line) from the muzzle at point A to its landing at point B. That certainty is the basis for the entire field of ballistics and for projectile-based weapons in warfare. Yet there are infinite possible paths connecting those two points, A and B (some of which are shown with dotted lines). So why does the cannonball always choose that one particular path? What is so special about it? It is easy to answer that now: along that path the "action" has the minimum value! Every other path would lead to a higher value of the action as defined above, just as was the case with our financial example. So this is how the *principle of least action* plays out: At each point along the trajectory, take the kinetic energy (expenses) and potential energy (bank-balance), find their difference, and average that over all the points on the

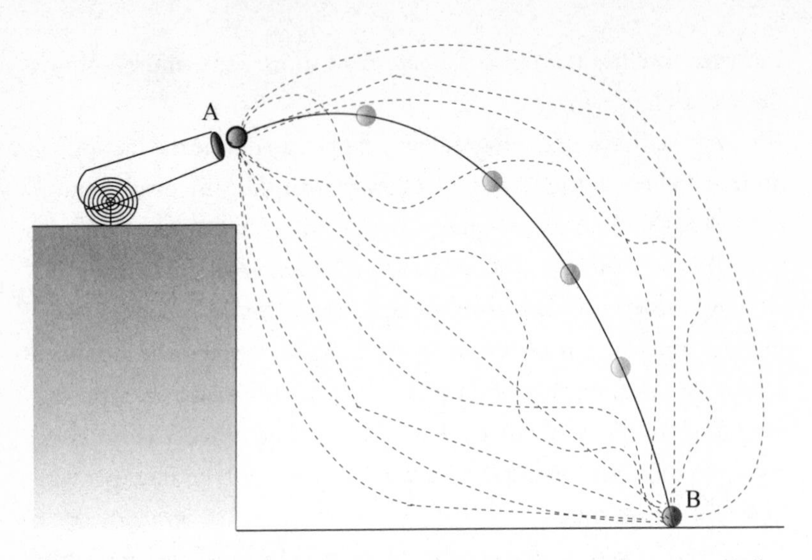

Figure 4.4 The solid line shows the actual path followed by the cannonball—the action has the minimum value along this path. Some of the infinite other possible paths are shown as dotted lines. They are not taken because the action would be higher along those paths.

path; the true path is the path that gives the smallest value for that average.

Now we have a simple explanation for *why* things move the way they do in the universe—everything is simply following the path of least action! And it all makes perfect sense once we identify *energy as being nature's equivalent of money* and accept that nature is thrifty in its use. By always following the path of least action, the universe manages to hold on to as big an average "bank balance" of unused energy as possible for as long as possible. That should also make sense of why we define action the way we do: how much action (in the everyday sense) is taking place can be gauged by the average energy invested in motion and subtracting of the average energy left in storage.

It is absolutely amazing that almost everything in the universe naturally charts out a trajectory that minimizes the action—as if every object out there, animate or inanimate, has an uncanny ability to make some conscious choices about the path it follows. Not only that, since the averages along the *entire* paths are involved in such choices, it seems that every moving object seems to have teleological information about all the paths *in advance* even before it starts to move! Well, I can only assure you here that there are no conscious or preordained choices involved. But there is indeed no satisfactory explanation for this in classical physics. An explanation can be found in Feynman's Path Integral formulation of quantum mechanics that we will describe later in Chapter 20. Yet even without delving into that next level of "Why?" we already have something extremely powerful in the *principle of action minimization* because it gives us a simple guiding principle for finding the correct dynamical laws about how things move and evolve in time in the universe—physicists use it all the time; really, we would be lost without it.

Nature's demands for minimum action translate to something we can all relate to in our "natural" predispositions: Reduce the kinetic energy (cut back on necessary motion) and increase potential energy (store up energy) as much as possible over time, because just like in nature, we humans have a natural tendency to minimize action all the time. In fact, we absolutely love minimizing action, like sitting on a couch with a bag of chips—our kinetic energy is as small as possible (we are not moving!), and the bag of chips we are munching on is increasing our potential energy in the form of stored, unused fat in the body. The other denizens of this world take this action minimization principle even more seriously; most animals don't move around much unless they have to (a lion sleeps and rests most of the day), and when they move at all, it is mostly because they are compelled to increase their potential energy in the form of food. So couch potatoes among us can't be held at fault, they are just following one of the most fundamental laws of the universe: action needs to be minimized! Lying on the couch with beer and snacks, remote in hand, leads to a minimum action on our part, and that is exactly what the universe demands.

Every day, the principle of least action influences our behavior and movements and those of all the creatures around us—in ways that seem to arise from instincts and intelligence. For example, the shortest path between two points implied in the phrase "As the crow flies" is not a choice fundamentally initiated by avian intelligence, it is just the principle of minimum action at work—the same principle that sets the path for inanimate objects, as well. Very often, the paths of minimal action manifest themselves as the paths of least resistance. Doesn't that also sound familiar? Look at how water flows or a ball rolls about, they always seek out the easiest paths, the ones going downward and with the least number of hurdles. Don't we all do the same thing? Leaving aside other motives like exercise or sport, we instinctively find the easiest and the shortest path. If we need to go from point A to point B, and we have the choice of a long windy dirt road that goes through rough terrain or a smooth paved road, which would we rather take? When walking through the manicured quads and lawns of a university or a corporate campus on a sunny day, we need to draw on our reserves of civilized upbringing to suppress our natural inclination to cut across the grass in a shorter diagonal, instead of plodding along the nicely paved but much longer path around the grass. Because you see, on the level ground of a quad, the potential energy does not vary much and is essentially constant, so minimizing action boils down to minimizing the energy invested in motion, and that would be along the shortest path—the diagonal straight line across it.

Now, we might very well wonder why not just make the kinetic energy zero, which is as low as it can get. Well, we can, but only if we do not need to go anywhere. But if we want to get places, we have to move, so our kinetic energy cannot remain completely zero. Once we do start moving, it is then all about striking the right balance between cutting down on the energy of motion and holding on to our potential energy as we move along the path we take.

Speaking of energy, there is actually a somewhat simpler principle that applies in numerous situations that simply states that physical systems prefer the state of lowest energy available. In quantum mechanics, we call that state of lowest energy the *ground state* (in analogy with the ground floor being the lowest floor in a building). If a system is constrained such that its amount of energy cannot be transformed into heat (which happens to be just disordered energy as seen in Chapter 3 when we talked about the second law of thermodynamics), the system will eventually settle into the minimum possible energy. On the other hand, if we allow some of the energy to be transformed into heat, it will still try to cut down on all its potential energy to maximize the loss into heat. In either case, the upshot is that nature usually gravitates to states of lower energy.

Minimize energy, minimize action! Ah, what a justification for laziness! We humans are lazy by nature, and it turns out that so is nature. Most of us probably would not be doing much with our time and our lives without some external incentive and a fear of the consequences of not doing anything. We would just hang out in our ground state of sleeping and lounging around watching television or hanging out with friends in some pleasant surroundings. We even have a name for the ground state—we call it "vacation," and that is what we look forward to for the rest of the year. If we had a choice, life would be just one endless vacation—a lifetime in the ground state.[8]

[8] Some folks are very physically active during their vacations—well, even in nature there are occasionally strange scenarios where action is maximized rather than minimized.

Even marriage and stable relationships are in many ways the paths of least resistance and least action through life. There is a physical law that states that when two objects attract each other, then the closer they get, the lower is their total energy; likewise, when two objects repel each other, their combined total energy is lowered the farther apart they are. We will have occasion to talk more about this later in Chapter 10. But for now, just from the point of view of minimizing energy, it makes absolute sense why we thirst for a good stable relationship or a marriage with abundant mutual attraction, where two people just get along and never waste an ounce of energy in arguing, fighting, or domestic violence. No one likes to remain single forever, because single people have to invest a lot of energy and effort in trying to find dates and someone to be with, someone to hang out with—that is too much action and too much energy wasted. So people get married and try to settle into stable relationships to minimize action and energy. And if we were to go by the confidential and resigned testimony of many married men, it seems marriage significantly minimizes the "action" in the bedroom as well!

Believe it or not, it so turns out that the universe is *not only lazy but also seems to be in a tremendous hurry*, conveniently providing a natural justification for the endemic impatience of our species. It is not a joke—there really is a *principle of least time*. Among other things, it explains why light bends when moving from one transparent medium (like air) into another transparent medium (like water). *Refraction,* as it is called, is the reason why things under water always appear to be closer to the surface than they really are; ask anyone who's gone spear fishing. It is also why a drinking straw in a glass of water looks like it is bent at the surface. It happens because light follows the path that takes the least time, and since light travels slower in water than in air, it turns out that the shortest time of travel between two points—from under water, where the object is, to above water, where the eyes are—is not along the shortest path distance-wise, which would be the straight line joining the points. Instead, it is along an appropriately bent path, as shown in Figure 4.5. An analogy will help. You are hunched over a map trying to figure out

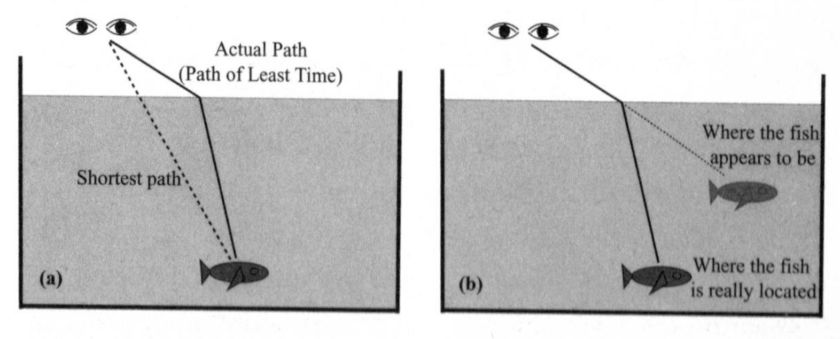

Figure 4.5 When light travels from one medium (water) to another medium (air) where its speed is different, it bends at the surface in order to take the path of least time of travel. (b) Since our eyes do not know that the light has bent, we interpret the origin of the light as extrapolated back along a straight line, so that an underwater object (like a fish) seen from above seems to be in a different position from where it actually is.

your route for a road trip. If you are trying to minimize your travel time, you would certainly take the beltway or the bypass around a large city rather than taking the direct shorter route straight through the city because you know it is not just the distance, it is also about how fast you are moving—so the most direct and straight path is not always the path of least time. These days, GPS and online maps give us the choice of what we want to minimize in our trip—time of travel, distance, or tolls—and they might all correspond to very different routes. In the case of propagation of light, the choice is to minimize the time of travel. Just like in a road trip, the path that minimizes the time of travel of light—from a point within water to a point outside—is not the straight-line path of shortest distance. That actual (bent) path can be found mathematically by considering the different velocities of light in air and water.

This notion will certainly be welcomed with an "I told you so" in these days of instant gratification. If the fastest entity in the universe—light—is in such a hurry, then perhaps we humans can be forgiven for our rush to get everything and everywhere instantaneously. Electronic devices and internet have completely spoiled us; we expect now to get everything instantaneously—to have immediate access to music and videos on YouTube, or information about anything at all with a click of a button on Wikipedia or the web, to talk and interact with anyone anytime with cell phones, emails, and Twitter. We are getting an opportunity like never before in human history to indulge our natural impatience. Thoughtful reflection and patience are fast becoming lost virtues. We could almost say that our impatience, besides our laziness, has been the other driving force for much of our technology—our need to travel faster, need to communicate faster, need to cook faster, do everything faster, faster, and ever faster. The irony is that the upshot of it all has been that we as a culture have less time than ever before, because expectations around us have changed as well, demanding speedier response from us on everything.

The good news is that if you happen to prefer the path of least action, the way of minimum resistance, the states of lowest energy, and are chronically impatient to do it all in the shortest time possible, well, there is nothing wrong with you. You are in the right universe.

CHAPTER 5
GLOBAL EFFECTS OF POTENTIAL GRADIENTS

"Globalization" is a favorite buzzword that raises all kinds of passions. It affects us all in every way from finance to culture, yet the process often seems too complex to comprehend. But much of it can be understood with a simple physics principle, potential gradient, *which occurs whenever there is any sort of imbalance between two regions.*

This recent global financial crisis has been analyzed to death. I have heard of every possible cause cited and blamed with passion and with vehemence—that is, every cause except the right one. I have read the news, and I have listened to the experts, and I realize that the favorite scapegoat has been that legalized con game of subprime loans in the housing industry. Were subprime loans instrumental? Absolutely. But was it the root cause of the whole thing? Definitely not—it was actually more of a means rather than the underlying cause. The real cause had already been brewing for a few decades in a variety of fundamental shifts and changes in the international economic and political scene. Although the actual dynamics of how things played out in the real world of finance, politics, and greed is monumentally complicated, behind it all there is a very simple principle at work—straight out of an introductory textbook of physics.

But to understand that root cause of it all, we need to take a closer look at what has been going on steadily but surely in the backdrop: Globalization has been the buzz for the last couple of decades. But globalization, as far as international trade goes, had happened and even boomed several times before in the course of human history, albeit on smaller scales, and most remarkably in the colonial years leading up to the First World War. But our collective memories are notoriously short-spanned and do not go back more than a generation or two, so by *globalization*, here, I refer only to the relatively recent developments of global economics and geopolitics that have happened since the 1980s. Nevertheless, all that is said in this chapter can be applied to more historical scenarios as well.

Anyway, have you noticed how our attitude toward this recent bout of globalization has shifted throughout the years? When it first began to get our attention as a significant phenomenon, it was universally regarded as the greatest thing to have happened since sliced bread. "Global this, global that"—we just could not get enough. Well not so much anymore, because lately globalization has been showing a sinister dark side most of us did not expect. It has become a divisive issue—emotions and passions run pretty high on all sides. While the positive effects were eagerly embraced and anticipated, the negative impacts came as a surprise to most. They still continue to surprise and confuse—because these days the world is in flux, changing faster than we can keep up with, driven by forces hard to comprehend—and it is easy to lose orientation and perspective caught up in it, as we all are, like flotsam on a mad raging river. Before getting into the simple physics behind this complex process of globalization that affects us all, it will help to take a couple of snapshots of its real-world effects from a few different perspectives.

What does globalization mean for developed first-world nations? First the good: Prices have dropped, and consumer products have become cheaper, because cheap labor overseas is bringing production costs down. Homegrown companies have a huge market spanning the globe. Jobs that you consider not good enough for you, illegal foreign workers rush into for really low wages, much to the delight of off-the-books cash-paying employers. The world has become a familiar place—the icons and the products that you grew up with now have their subsidiaries and franchises everywhere you go on the planet; no more culinary surprises for you when you travel—McDonalds and Starbucks are waiting for you at your destination. Really, if you want to find the center of any major city in the world, you don't need a map, just seek out the golden arches.

Then the bad: With outsourcing, jobs have been steadily disappearing. Low prices are great, but when you do not have a paycheck to buy anything with, it matters little. Then when you are finally willing to take any job that you can, you find that they have all been taken by illegal immigrants who are willing to work for pittances and under conditions that just do not seem to be worth it. Those franchises may be comfortingly familiar, but they are also steadily wiping out local eateries and coffee shops, homogenizing the world, and chipping away bit by bit at the charm of being in a new place.

What about the other end of the spectrum: What does globalization mean from the perspective of some of these rapidly developing Asian economies? First the good: Outsourcing is great; with a college degree it is now easy to find a job that pays decently. You have your choice of all the imported consumer goods that only the ultra-rich could afford in the past (if they were lucky enough to find them at the local smuggler's shop). Now, you and everyone you know can buy a car and a flat in a high-rise using the novelty of easily available credit, whereas before, you had to save for years to buy even a scooter, and half your money used to go for rent. You can't get enough of the American food chains—great service and food that seems so exotic.

And then the bad: You got a job, great, but now you work all day—in a private company that cares for little else besides profit—you are stressed all the time; you are in a city far from where you grew up—your old way of life and the sense of community are just memories now. Foreign cars are great, but the traffic jams and the pollution are ruining it all. And all those fancy goods you got by pretty well without not so long ago, you just can't seem to do without because everybody you know has them—so now your much bigger paycheck still cannot keep up with the expenses. And all that delicious fast food is thickening your arteries, and heart disease and obesity seemed to have tagged along with the outsourcing.

And there you have the great irony of globalization: exactly those things that made it seem such a great idea at first are also the ones now turning around and biting our behinds. How could something that started off so good end up screwing us so bad? Can we do anything about it? Can we reverse it? Should we reverse it? The answer to all of that lies in one physical concept: *Potential gradients*—the good, the bad, and the ugly of globalization are all about potential gradients and the *flow toward equilibration*.

Even if the term *potential gradients* may not be familiar to all, I am absolutely certain everyone is quite aware of the consequences of the *physics of potential gradients*, because we all know that water flows from high to low, that electric current flows from high voltage to lower voltage, that things fall from heights to depths, that heat flows from hot to cold (or else we would not need to insulate ourselves and our homes in the winter). All these things happen due to *potential gradients* (often also referred to as *potential differences*): They arise between two physically separated regions whenever there is a difference in the energy or the concentration of any substance of interest between the regions. The region with the higher amount is said to be at a higher potential and vice versa. For example, consider water behind a dam as shown in Figure 5.1. It is at a higher potential because there is a lot of it filled to a higher level relative to the bottom of the dam, so there is a potential difference as regards to the height of the water behind the dam compared to the water level below the dam. But here's the essential point about potential gradients: Unless there is a restoring mechanism (like a river flowing into the dam), the potential differences can be sustained only as long as the two regions remain isolated from each other. As soon as the regions are directly connected, there will be flow from the region of higher potential to the region of lower potential, just as water would flow out if a sluice were opened in the dam.

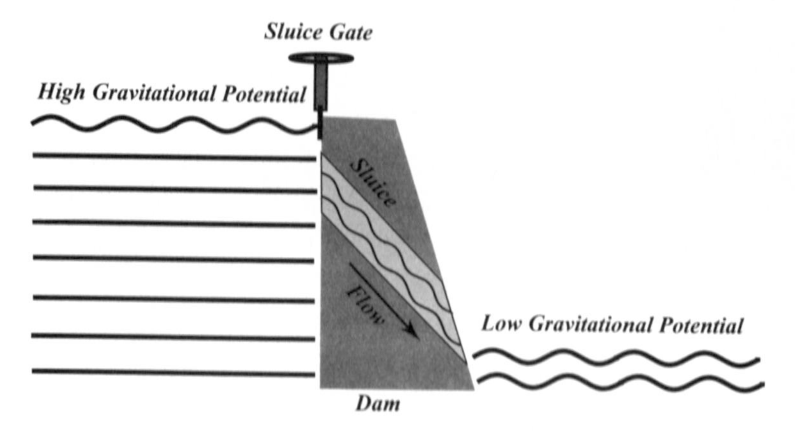

Figure 5.1 When the sluice gate is opened, water contained by the dam hurtles down the sluice from a region of high gravitational potential (greater height) behind the dam to a region of low gravitational potential (lower height) at the bottom of the dam.

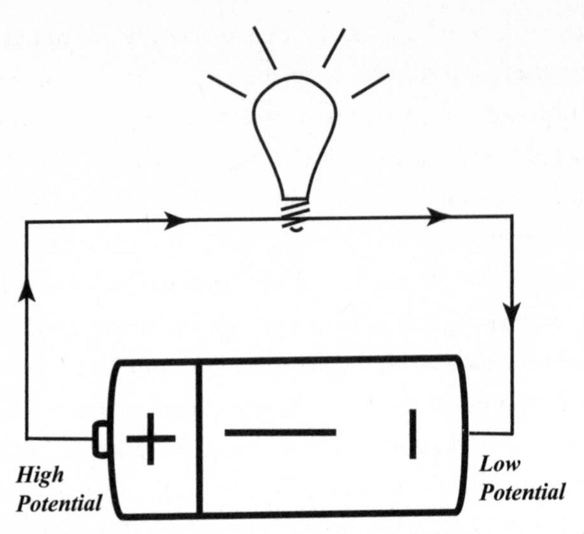

Figure 5.2 The arrows show the direction of flow of electric current through a wire connecting the terminals of a battery. The flow is always from higher potential (+) to lower potential (–). That flow is harnessed for useful work, like lighting up a light bulb.

Every kind of flow is driven by some form of potential difference. Because nature always tries to balance things out, nature seeks equilibrium and nature, left to itself, is the true great leveler. We have to work against nature to create and maintain height differences, voltage biases, temperature gradients by creating artificial barriers, like by building a dam, by isolating the positive from the negative terminal in a battery (they do that in the factory), or by packing insulation into the walls of houses. In the language of physics, all these are *potential barriers* that we put in place to maintain potential gradients—gravitational, electrical, or thermal. But why call it "potential"? Because letting the barriers down leads to flow—of water, electricity, or heat—which has the "potential" or capacity to do work, like running a hydroelectric turbine, or powering appliances around the house, or turning a windmill by thermal gradients in the atmosphere.

That is exactly what globalization is doing in the socioeconomic arena: It is removing the artificial man-made human potential barriers of national boundaries, leading to flow of people, money, jobs, culture—and everything else of human value and interest—among different nations and regions of the world. Everything is in flux right now, trying to seek out a new equilibrium, assuming that any exists. The blessings and frustrations of globalization—both can be well understood from this process of equilibration. Pre-globalization, all kinds of human potential differences—economic, demographic, sociocultural, and geopolitical—had built up over time due to barriers of national boundaries, political isolationism, iron curtain, travel restrictions, communication gap (pre-internet and pre-Skype), geographical separation, different ideologies, varied population densities, and the list goes on. Most of those barriers still exist, but globalization has eroded many of them and continues to chip away at all of those barriers.

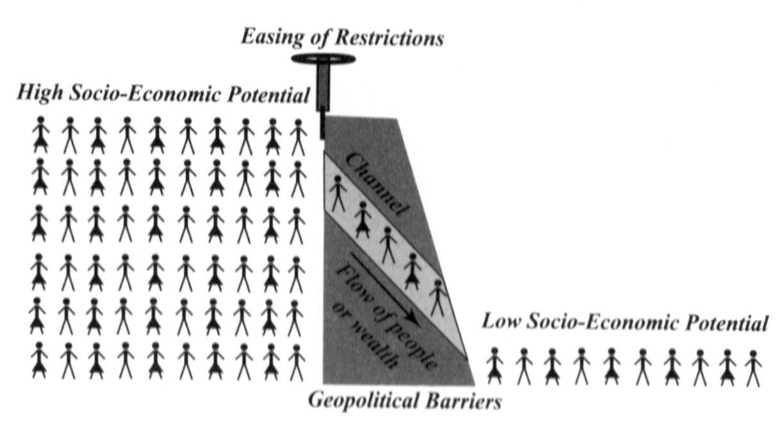

Figure 5.3 When there is socioeconomic disparity in wealth, population, freedom, or other such variables among two regions separated by a "dam" of geopolitical barriers, then human potential differences build up. If these barriers are removed or reduced, there will be flow of people, wealth, and culture. (Note: The stick figures are meant to represent any relevant human socioeconomic factor, not only population density.)

Cultures are becoming mixed, and their distinctive features are either being lost as foreign ones overwhelm them or are being drastically transformed as they blend to create new hybrid ones. Occasionally, the end result is something new and more exciting—like fusion cuisine—and is a cause for rejoicing, but more often than not, age-old traditions are wiped out forever and leave a sense of loss in many. One can understand the chagrin of people as they see many cherished things they grew up with slowly vanishing in front of their eyes. But there is little that can be done, because with the potential barriers down, the flow is impossible to stop, and even if it were possible, *nobody really wishes it to stop altogether*, only the parts they don't like! But we cannot stop it partially—it is all or nothing. So when my French friend complains to me about McDonald's in the center of French towns, he cannot really reject such franchises without rejecting much of America as well, and deep down he knows that is not possible, neither for him nor his nation.

Many of the economic woes in the West today can be directly attributed to the equilibrating tendencies of globalization. In the decades after the Second World War, an increasingly widening economic potential difference opened up—in living conditions, average income, lifestyle, and general expectations in life—between the developed countries mostly concentrated in the West and the emerging developing countries elsewhere in the rest of the world, particularly China and India. Those potential gradients grew greater and steadily more pronounced behind the barriers imposed by the Cold War and protective economic policies. Then as the Cold War drew to an end in the early nineties, coincidentally or not, many of those barriers diminished and some disappeared altogether. China and India, for example, let go of some of their protectionist policies and opened up their markets. Just like when a dam breaks and the water rushes through, it did not take long for a massive socioeconomic flux to burst through, with large companies pouncing to take advantage of the built-up economic potential gradient. Outsourcing got underway with a vengeance; first manufacturing got shipped out to countries like China with an abundance of cheap labor, and then the service sector started outsourcing to mushrooming call-centers in countries like India where a large English-speaking, job-hungry, young population was ready and willing to pick up the phone and deal with irate customers for a fraction of the salary in the West.

Corporations and banks, and the people involved with them, used this financial torrent released by economic potential gradient to generate more money for themselves as they saved operating costs by shipping jobs overseas where labor costs are lower and bolstered profit by selling cheaper products at higher margins back where the labor costs are high—pretty much the same way you can generate energy from the flow of water from high to low, or electricity from high voltage to low voltage. Caught up in the rush to make more money using the opportunities so released, people overlooked a few critical things about removing potential barriers in any system, be it be in physics, nature, or the human landscape. The first crucial thing is that any flow started by breaching potential barriers *cannot last forever*; just as a battery eventually drains, eventually the potential difference will disappear, and the torrent will become a trickle and perhaps vanish altogether, unless there is a way to restore the potential difference again. The next quite obvious consequence of the flow is that *regions of high potential get diminished* eventually, meaning richer regions will get poorer. But perhaps the most disconcerting effect is that the potential barriers broken down by globalization released a *whole menagerie of torrents, many of them unintended and unwanted*, affecting aspects of our lives that we thought were sheltered and secure. For example, jobs started flowing out as well, hitting some economies really hard.

Still, people continue to blame all the wrong things for the economic crisis—subprime loans in the housing market, wild speculations on Wall Street, bad governmental policies. But in reality, all these are simply the instruments of the crisis; they are not the root cause. Human greed and human behavior has always been the same and will remain so in the foreseeable future. People *en masse* have not changed and will not change; people simply adjust their behavior and react to changing circumstances but always with the same predictable basic instincts—the perennial interplay of greed and fear. It is no use blaming corporations for trying to make more money—that is what they are supposed to do—or ordinary people for trying to buy houses more expensive than they can afford and living beyond their means; it is just human nature. The real culprit in the economic crisis is the breakdown of economic potential barriers due to the post–Cold War globalization. The playing field of companies and corporations suddenly broadened and expanded immensely, helped also coincidentally by the simultaneous breakdown of information barriers due to the internet and the resulting growth of online and computer-based trading. With stocks, bonds, options, and derivatives traded across the world without boundaries and without restrictions, financial instruments have become a genie out of a bottle with nobody or no nation with complete control. I doubt even the top executives of any company have a complete and clear picture of what is really going on: How could they? It is more than any one human being can comprehend. In such a free, unrestricted environment, speculations are bound to get wilder and values of derivative financial products bound to lose any semblance of grounding in concrete value. Just like the turbulence in the mad torrents bursting out from a breached dam, all the national economies have burst forth to merge into a global maelstrom, and the scary thing is that now, if one big economy is in trouble, the whole world is in trouble, because everything is interconnected.

Finance and economics are not the only arenas affected by the relaxation of the human potential barriers over the last few decades. It has also directly affected the physical flow of people across the world. One visible and troublesome example of this is the issue of illegal immigration facing much of Europe and America, and this phenomenon can be quite easily understood in terms of economic potential differences. When we have a rich country next to one full of extremely poor people with a hopeless future, what can we expect? With such a huge socioeconomic potential difference in such close proximity, there will be flow of people unless there is some seriously high and insurmountable barrier, physical, governmental, or military. Stopping that flow is as hard as stopping a river from overflowing its banks—the rising poverty, just like rising water, will find the smallest opening and force its way through to the other side. Look, even the brutal potential barrier of the iron curtain was barely enough to keep people artificially separated among parts of Europe where the socioeconomic differences were nowhere near as stark as those that exist between say Europe and sub-Saharan Africa.

Here's the unspoken dilemma of the developed countries: They have walked themselves back into a corner by promulgating high moral ideals of human rights as a shocked backlash against the genocidal horrors of the twentieth century. The legal and moral authority of these countries is based upon respecting human life and giving a fair and equal chance to everyone no matter where they are from. As a result, when illegal immigrants show up on European shores with no papers of any kind, officials often have few options besides just turning them loose. A solution under these circumstances is physically impossible. We cannot destroy all the barriers and expect the flow to stop. The flow will continue until there is no potential difference. The stronger the potential gradient, the stronger the barrier needs to be. If people are willing to brave death on the high seas on frail rafts, they must be desperate indeed, underscoring the high potential difference that exists in human terms. They would hardly be dissuaded by a bit of jail time or being sent back to where they came from—tactics currently used in Europe and in the United States. It is like trying to throw up paper embankments to keep out a tsunami. Yet, within the current legal and moral structures, puny measures are all that are available. This is not about taking sides one way or the other on this very tricky issue or about defining what is right or wrong. This is simply pointing out the reality as dictated by the physics of potential gradients: the larger the potential difference, the stronger the potential barriers needed to prevent flow. What such barriers would be, well, your guess is as good as mine.

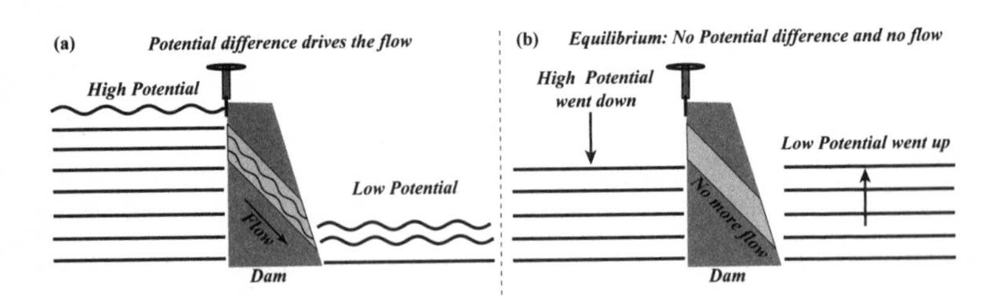

Figure 5.4 (a) As long as there is a potential difference, the flow continues. (b) The end result of flow is equilibrium when there is no more potential difference and the flow stops. But this means the high potential region went down, and the low potential region went up.

Since this is such a sensitive issue, a slight digression is called for to make a distinction from the almost antipodal scenario of legal immigration—in fact, the emergence of the United States of America as a global superpower is a great example of how legal immigration can actually be a shot in the arm for the economy and the general well-being and vitality of a nation. That's because legal immigrants can immediately feel a vested interest in their new nation even by such simple acts as paying taxes, and in return they have direct and legal access to all the benefits offered by their adopted country. On the other hand, for illegal immigrants, the dice is loaded from the start, since they have to stay under the radar; they also weaken the system by burdening it without contributing any taxes toward it. Therefore, it should be understood that all the problematic aspects mentioned here on this issue pertain only to illegal immigration, although for sure, similar socioeconomic potential differences drive legal immigration as well.

Anyway, to return to the physics of it, the end result of all flow is *equilibration*—which might sound like a very nice thing as in "everything is approaching calm again!"—but the trouble with equilibration is that it inevitably means that the regions of high potential are lowered, as we can see in Figure 5.4, and that is already happening. For example, if the flow of illegal immigrants continues, it will bring down the wealth and prosperity of the destination nations as their social systems get overburdened. So eventually one of two things will happen: either the flow will continue until the developed countries are reduced to lower socioeconomic levels, reducing the socioeconomic potential differences that are driving the influx of immigrants, or the desperation in the developed countries at losing their way of life will reach a boiling point when they start enacting stronger and inevitably brutal measures that transgress the current moral and legal fabric. I personally believe it will be the latter option, because I can already see more radical viewpoints and measures gaining currency in places like the American Southwest or in parts of Europe. People stay civilized only as long as their personal interests and security are not endangered. It would be a capital mistake to think that we as a species are beyond acts of violence and war—after all, we are only seventy years away from the deadliest war in history, in which 45 million people died in conflicts among "very civilized" countries. Our brains and instincts do not evolve that fast. If we have avoided any major conflict since, it is because people in the strongest nations have been generally well off and happy, and the memory of that last devastating war lingered, but most importantly because nuclear weapons changed the rules by introducing Mutually Assured Destruction (MAD), where there can be no winners. Actually, there are some elements of ironic truth in the cynical point of view that nuclear weapons should get the real credit and several Nobel Peace Prizes to boot for avoiding a Third World War—at least so far.

The principle of how potential differences drive flow is universal and supersedes human emotions and wistful thoughts. Once we remove the barriers, then there will be consequences, some to our liking, some not; for example, outsourcing saves money but bleeds jobs as well. In a global environment, no matter how outraged local workers are, paying someone ten times more for a job here simply cannot be sustained when thousands elsewhere on the planet are happy to do it for ten times less because, for them, even that is far better than the alternatives. In an open global market, we all have to compete harder to survive.

In nature, the sun powers the atmosphere to create the precipitation that makes the rivers flow to sustain the potential difference across a dam. But in the geopolitical and socioeconomic realm, we can't count on the sun to maintain the potential differences.

Therefore, there are only two possible outcomes: reinstate political and economic barriers *or* come to terms with equilibration where the wealthier countries will have to lower expectations. But as we all know, both options are easier said than done; nature also teaches us that building up potential differences is a much harder and longer process than making use of potential differences.

CHAPTER 6
COPING WITH IT ALL BY SCALING AND RENORMALIZATION

Scale, as in magnitude—tangible or hidden—has a profound influence on our worldview as well as on the relevance of nature's laws. Appearances and behavior can be completely different at different scales. But certain things remain the same, or scale-invariant. What changes and what doesn't have close parallels with how we come to terms with all the inevitable changes and events that happen in our lives.

You may never have given it much thought, but the physical, emotional, and even moral landscapes that define the way we all experience life are inherently defined by the concept of *scale*. Scale is also at the very foundation of how science is done in practice, particularly the physical sciences. Physics is often called an "exact" science, giving the impression that everything that physicists work with and every result obtained is exact—that every computation and every measurement is done to infinite precision and accuracy. But the reality is quite different, because for practical applications, we never need to know anything to infinite precision, even if we have the means to do so. So when we toss around the word "exact" in the physical sciences, what we really mean is that things are exact up to the relevant "scale," which just happens to be more precise relative to other less exact human endeavors.

In practice, science and technology are based on working within the tolerances and approximations defined by the scale of interest. What does that mean? Well, in the physical world, knowledge of a system can be at different levels, or scales. For instance, if we want to know how a baseball moves when we throw it, we have no need to know the molecular composition of the baseball nor do we need to know how the baseball field is being carried around the sun by the revolution of the earth. Physical theories are often applied only at a certain specific scale, which typically does not necessitate knowledge of what is happening at another scale of the system. A scientific description of a baseball can be done at many scales: There is the scale of the molecules inside the baseball; there is the scale of the stitches and the material that make up the baseball; there is the scale of the ball, bat, and the players; and above that the scale of the field where the game is played; and beyond that the scale of the motion of the entire earth that is dragging the ball with it; and so on it goes. But we can ignore the dynamics at all the other scales, except the one that is relevant: the scale of the ball, bat, and the players. And because we ignore what is happening at all those other scales, our calculations would not be infinitely precise or exact, but it is sufficient for the practical purposes of predicting how the ball will move within the limits of human perception—anything more would be pointless for that goal.

If we were to take the classical worldview of Newtonian physics, that the entire universe is some sort of clockwork mechanism, this matter of scales would simply translate to a regression of exactly the same laws of physics applied at smaller or larger scales. For example, the laws of motion that govern the baseball would be exactly the same as those that determine the motion of the earth around the sun, just on different size scales. The motion of the earth would simply add a very tiny correction to the motion of the baseball, which we can comfortably ignore because various other uncontrolled factors like wind will have much stronger effects. In any case, this applicability of the same basic laws across different scales is true for a wide range of physical scales, including the two just mentioned (always, of course, up to the relevant precision!). But what quantum mechanics and relativity have shown in modern physics is that even the very laws of physics could change at a different scale! Scale is therefore not always just a matter of precision; it can fundamentally change our worldview.

Take nanotechnology, for instance; it is the buzz these days, and it is really all about scales. Nanometer simply means one-billionth of a meter, so that we can fit a billion nanometer-sized objects side by side on a meter stick. That is obviously a scale of things completely hidden from that of the world we operate in on a daily basis. "Small" in this case happens to be very different. At that nanoscale, we can no longer use classical physics that we use at the scale of our everyday life, like in describing the motion of a baseball. At the nanoscale, we need quantum mechanics, because strange things can happen there that we do not see in everyday life—particles can go through barriers, they can simultaneously exist in many places, or can be entangled with other particles far away. But the key point here is that the nanoscale quantum world is not in some far away and alien universe disjointed from us—that quantum world is right here within you and me and within everything in this universe. Yet, until recently, we had been getting by quite fine without knowing that this magical inner quantum world even exists. We still can ignore it completely in our day-to-day life. It's a stark reminder of how scales separate us and how the concept of scale can fundamentally alter our vision and understanding.

As an example with more direct implications for us, consider the world of microbes and germs. Here, we can all appreciate how the scale of things can completely hide from us deeply relevant entities, which nevertheless have enormous impact on our lives. Until the microscope was invented, the tiny size-scale of germs kept their nefarious work hidden from a long-suffering humanity who could only attribute the scourge of diseases to the punishments of a fickle divinity for most of civilized history.

The deep relevance of scale is therefore quickly apparent in the natural world as we face the multitiered structure of the physical universe that extends from the realm of elementary particles at sizes much smaller than a nanometer to the intergalactic spaces that stretch millions of light-years.[9] We might feel content to let science deal with all that since we are all comfortably ensconced in our one particular relevant scale of things—the human scale—and assume that we can ignore all other scales and not even bother with the very concept of scale itself. But we can't! In truth, scale defines every aspect of our lives in the most fundamental ways—and even more broadly and deeply than in the natural world, as we will see in the rest of this chapter. Becoming aware of exactly how brings to question many of the comfortable beliefs that we take for granted about ourselves, individually and as a species.

The most obvious implication of scale in our lives is the most literal one: We humans exist on a scale defined by our average size. We are so used to it that we never think about it. But just imagine how our lives would be if we were the size of ants; we would exist in a world of unbelievable monsters, we would even be terrorized by spiders and frogs (some of us already are!), and we would have a hard time comprehending an elephant or a blue whale. We can

[9] One light-year is the distance light travels in a year in vacuum and is approximately 5.9 trillion miles! It is used as a unit of astronomical *distance* (not time, despite the confusing presence of the word "year" in it).

get some idea from the movie, *Honey, I Shrunk the Kids!* and its sequel. Even small changes in scale can significantly affect our lives; anyone afflicted with dwarfism or gigantism would tell you that being adults who happen to be just a bit smaller or larger than the average makes everyday life a challenge, enough to justify some reality TV shows to document those challenges.

In fact, one could seriously question whether we would ever have been able to develop the technology that we have today if we were the size of ants and everything else on the earth remained the same—even if we somehow had the same intelligence. It certainly would have been a much bigger challenge. Imagine building a bridge over a major river like the Hudson or the Mississippi if we were the size of ants, or navigating the oceans when each ordinary wave is a hundred times bigger than us. At the other extreme, if we were super large, each of us the size of a whale, would we have been able to grow enough to feed us all, would agriculture even have developed, would we even have noticed those annual regenerations in tiny little grains of wheat, corn, and rice? And without agriculture, civilization simply would not have started, because like other animals, we would be foraging for food most of our waking hours. I doubt it is a coincidence that in the size scale of living things on the planet, we happen to be at the size scale that we are, which happens to be quite close to that of the larger creatures on the planet. For sure, we are not as large as a whale, but certainly the difference in scale between a human and the largest land animal, the elephant, is negligible compared to the size difference between us and bacteria or even typical insects. It would be worth a serious research study and few doctoral dissertations to see how our natural size, relative to everything else on the planet, has been instrumental in defining our culture, civilization, and intellectual growth.

Our natural size scale influences many of our perspectives, as well. If we were 100 meters tall, climbing mountains would lose much of their challenge; however, if we were the size of ants, the peak of Everest might as well be on Mars—although climbing it would be easier in a way, since surfaces that seem smooth and sheer to us would be craggy for ant-sized humans. But what is more relevant is that our size scale defines our attitude toward other creatures. People get all worked up about the thought of whales and tigers and lions becoming extinct, but the reality is that myriad species are going the way of the dodo every year. Yet even the most conscientious activists and conservationists would not be able to work up the same kind of despair and emotion for the fate of most of them. And to a great part, that is due to the simple fact that the bulk of those vanishing species happen to be little bugs or fish or frogs on a much smaller scale than our own. Celebrities rush to protect dogs and cats and publicly feel the pain of turkeys and chickens, but I doubt you would find any of them pushing the cause for some unknown tropical insect. Yet if we wish to be objective and fair, those bugs are equally important with unique genetic codes that are lost possibly forever once they become extinct. But that is the way it is; our sensibilities and morality are inextricably tied to our sense of scale. Most of us would be wracked with guilt if we ran over a dog or a cat with our car, but we would not think twice about squashing a spider with a bathroom slipper. Scale discrimination is inextricably built into our moral landscape.

A more abstract, but also more profound, impact of scaling is on our emotional and mental makeup, which has an inherent sense of scale built in and which is absolutely fundamental for our sanity and survival. Our mind automatically scales our perceptions of everyday issues, starting from our relationships with people to the problems in our lives. At the top of that scale of importance resides our own selves—we are constantly occupied with ourselves, and even the most trivial things that happen to us matter. Next are those closest to us—our immediate family and the very close friends who are the most important people in our lives—they dominate much of our thoughts; they exist at the next highest scale of our consciousness, so that almost everything they do, almost every day of their lives, is important and relevant to us. Then there are the not-so-close friends and relatives, who are important to us but not quite as much, and they exist on a lower scale of emotional attachment, and their activities are averaged over so that only the major and cumulative changes in their situation over a period of weeks come to our attention. Well, we can continue to add several more levels on this scale or hierarchy of people's relevance in our lives depending on how broad a circle of people we are tied to. But at the very bottom of that scale, there is the rest of the world, and here we are really averaging over thousands and millions of people sometimes over months and years, and only something on the scale of a national disaster, big enough to make it to the news, registers in our heads. This is not callousness; our brains have to do that. We simply cannot handle caring for everyone and everything at the same scale; our sanity demands scaling.

The importance of the appropriate scale is implicit in the all-too-common admonishment about "seeing the big picture" or "missing the forest for the trees." They are warnings to the errant novice and the seasoned veteran alike about the pitfalls of dwelling on the incorrect scale of things. Say your job is to make millions for your company. Then it would be a mistake to cut corners by using cheap stationery and office products that risk making a bad impression on your clients—you would be "penny smart, pound foolish"—the wrong scale of things altogether. Scale-confusion with potentially far-reaching consequences are manifest in the debate about global warming—a phenomenon marked by trends that happen over decades and centuries—and we risk missing the point altogether if we remain obsessed with the extremely cold winters in our vicinity in the last two years, like many otherwise quite intelligent people seem to be doing these days. Likewise, in the past, some notable ancient characters might have interpreted a devastating local flood as a global catastrophe, although that would be hard to do these days with continual international news coverage. The truth is that even today, there are many who completely believe that recent disasters like hurricanes and terrorist attacks are signs for the end of the world—why? Because in our own mental scale of things, we are indeed at the center of it all, and just as a dime close to the eye can eclipse the enormous but distant sun, any catastrophe that affects us directly looms large in our minds and mental projections, even though such events might not be so significant on the grand scale of things. And because of their impact on us, we mentally rescale them to have universal significance. A local catastrophe, in time and space thus gets interpreted all too often, heart and soul, as something of global and even eternal consequences. So we might have some blighter seriously believing that the escalation of gang violence in his neighborhood is a telltale sign of Nostradamus's predictions or of some impending Biblical apocalypse. When we encounter such beliefs, it might help to remind ourselves of the real scale of things, perhaps by reflecting on why we continued to believe until a few centuries ago that we are at the center of the universe and even believed the illusion that the moon is as big as the sun.

A personalized rescaling of things is not always bad, however; in fact, it is downright essential for survival. For most of us, our minds have a coping mechanism built in that is based on rescaling and renormalizing. You might have heard that one of the big survival advantages of humans as a species is our adaptability to a broad range of situations and scenarios. This conjures up images of people living in the burning desert, in the wet tropics, and in the frozen Arctic. Definitely there is that physical side of human adaptability. But more importantly, we humans can adapt emotionally and mentally to almost anything. When things are going well, every irritation seems like a big deal, and at those times, if we hear about someone else going through a very serious crisis, we speak out in amazement about how that someone else is able to deal with such a terrible situation, and we are absolutely convinced that we ourselves could never possibly deal with something like that. But then, when such crises do come into our lives, as they must surely do sooner or later, most people can and do cope with them. Our mind naturally rescales the issues in our lives, and suddenly the minor problems are averaged out of our thoughts, and the mind comes to terms with the big problems and starts to deal with them.

Let me paint the picture in more graphic terms. Say you have a decent job, a good family and children, and are living the average middle-class life somewhere. Traffic is an irritation you could do without, the children constantly clamoring for attention gets on your nerves, the spouse's occasional insensitivity to your particular preferences, like the toilet seat up or down or not sharing in the cooking or doing the dishes—all of these just add up every day to keep you perpetually dissatisfied while keeping you feeling completely overwhelmed with work and chores. You are absolutely convinced that you could not handle much more of anything. You often wonder how the woman, who lives two doors down, does it with five children of varying age, one with Down syndrome, and the husband wheelchair-bound with a major disability. Yet somehow she always seems to be able to keep a smile on her face whenever you meet her. But the good news is that you could do it, too. If, God forbid, tomorrow your spouse comes down with a serious illness that will require prolonged care and treatment, your income gets cut in half, and your responsibilities double, will you just drop everything and run? Most likely not. Sure, there will be a period of transition filled with panic and despair, but eventually you will adjust to the new reality and find that once again on a daily basis you are thinking only of the new "minor" irritations, like taking your spouse to the hospital during your lunch break, taking the bus instead of the car to save money on gas and tolls, living in a smaller house with one less bathroom. *It again feels like normal, the new normal. You have just rescaled and renormalized your life.*

Renormalization—now that also refers to certain profound and exciting developments in physics in the last century. The theory of renormalization, in all its glory and subtleties, is quite complex. But some of its key concepts contain interesting parallels with what we have been talking about. In particular, one remarkable feature shared by some of the most fundamental theories in physics is that as we change the scale at which we examine a system, the structure of the theory is a self-similar replica of itself at each new scale. An illustrative way to appreciate what that means is to consider a system of regularly arranged particles, shown as circles with crosses in Figure 6.1(a). Imagine that this is just a tiny section of a much larger grid of millions of particles arranged like this. Suppose we have a physical theory that describes the system at a scale where we keep track of each individual particle. We might then want to simplify things a bit and take blocks of four particles as our basic unit, which cuts down the total number of individual entities we need to keep track of in the system by a factor of four, as shown in Figure 6.1(b). We could keep doing this, taking larger and larger blocks as our fundamental entity, and as the blocks get larger, we are essentially increasing the scale at which we are examining the system. Now, it so happens that in some of the most important physical systems, the theories that describe the system at *each of those increasingly larger scales are identical* in form in all their relevant features and parameters!

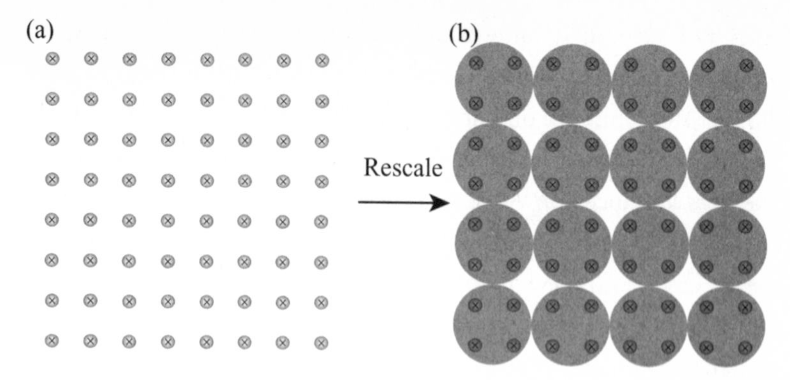

Figure 6.1 An illustration of rescaling: a system of millions of particles, marked by circles with crosses, are arranged in regular pattern, only a tiny section of which is shown in (a). If instead we grouped the particles together in blocks of four particles as shown in (b), then we would be looking at the system at a slightly larger scale as defined by the size of the enveloping circles. Ideas of renormalization imply that in many important physical systems in nature, the theory that describes (a) would be identical in form to the theory that describes (b) and so on for larger and larger scales.

There are direct human analogs of this. Say we are interested in developing a predictive theory of human behavior. We could look at every human being on the planet individually, but that could be a tough task, so we could simplify it a bit by looking at them as "blocks" of families, then simplify it further by looking at a slightly larger scale of neighborhoods and communities, and we could continue this way till we get to the scale of nations and countries. What we will find is that at every scale, the dynamics of human behavior is dictated by a certain set of rules and parameters that play out in a self-similar way: love, hate, hope, greed, self-interests, common causes of disputes, ambition, progress, knowledge, beliefs, faith, and so on. The details might vary significantly, but the determining parameters are essentially the same.

At a personal level, something like this plays out in the process of emotional renormalization that we were discussing earlier. Think of each of those particles with crosses as marking a source of irritation and trouble in our lives. When things are relatively alright, each of those tiny "crosses" impinge on our minds and perturbs us. But, when bigger problems descend on us (like the larger gray circles in Figure 6.1(b)), then we lose focus on those smaller irritations as we get occupied by these larger troubles. But, as with renormalization in physical systems, as we get used to this new order of things, our minds eventually settle into treating these new, bigger problems pretty much the same way as we did the smaller ones in the past. Of course, there is a limit to how much we can cope with, and our troubles could reach a certain scale at some point where we cannot renormalize anymore, and our life could just fall apart. Similar "breakdowns" can happen in nature, as well, at certain scales.

This natural rescaling or coping mechanism is not so uniformly present among us all, and for some, there are bouts of nervous breakdowns and hysteria, sometimes even in response to minor problems, like, "Oh my God! My cell phone stopped working! I don't know what to do anymore!" or "Oh no! I have a pimple on my cheek, my life is over!" A lot of spoiled teenagers and pampered celebrities come to mind. At the other extreme, we have the stoics and folks of heroic nature who seem unfazed by the worst possible personal disasters and take it all in stride. But most of us are somewhere in-between.

In a typical life, we do not even need any serious and sudden disaster to bring us face-to-face with renormalization—aging and the passage of time naturally imposes renormalization on us! The troubles of youth very often seem simple as most people face the realities of later life, yet those youthful troubles are serious enough while going through them, because that is the normal at that point in life. They only seem simpler and easier to deal with relative to the new normal of a more complicated later stage in life.

As with each individual life, there is ongoing rescaling and renormalization with every new generation. Here we are at the dawn of the twenty-first century, and we take so many things for granted—cars, air travel, computers, internet, cell phones, movies, excellent indoor plumbing, and a million other amenities unimaginable even a century ago. A middle-class family in a developed nation is living a more luxurious life than Louis XIV or any other great emperor or sultan of the past. So it would seem impossible for you or me to be able to live without all of those things we take so much for granted. But imagine you were born in the Middle Ages. You would have grown up with none of this, and you would have gone through life in the environment you were born into, taking that to be the *normal* with no knowledge of what you were missing. Billions of people have lived and died just like that. Forget the past—billions are doing that right now in less favored situations in less developed parts of the world. But the point is we normalize and scale ourselves to the situations we are born into, and we naturally adjust to the quality of life we are faced with, particularly when we have no knowledge of any other. In the sense of scaling in physics, it is as if we exist at a certain scale and are never aware of any other scale than that. Even literally speaking, we existed just like that until the last couple of centuries, with no awareness of atomic and subatomic scales or the stellar and the galactic scales, comfortable in the absolute belief that we were at the center of the universe and were God's only concern.

It is quite the same with nations and countries. In the absence of a big threat, we find them fighting wars over relatively minor issues with people who are in most ways quite similar. Then let there be a threat from a bigger foreign power that is a threat to both, and suddenly the small differences are rescaled away. You would be amazed at how minor some of those reasons could be; millions of Europeans died in the trenches in the First World War, in part because some member of outdated royalty got shot in the Balkans. The same European nations are now united in the European Union to compete with the socioeconomic challenges of the world. In the Crusades, a lot of the warring kingdoms of Europe put their differences away to unite in a common cause of protecting themselves against a powerful common threat. In the Sepoy Mutiny of 1857 in India, Hindus and Muslims, traditionally on opposing sides, were fighting side by side against the common foreign invader, the British. History is chock-full of such strange bedfellows in the face of common threats and shared adversities. I can promise you that if we get invaded by aliens tomorrow, then in the face of that threat on the scale of the human race, we will somehow average out all our differences among ourselves and remember only that we are all humans against those hideous aliens. Something quite like that plays out in a set of alternate history novels by Harry Turtledove, in which the world gets invaded by aliens in the middle of the Second World War.

We started this chapter talking about spatial scales (matters of size), and we will end it with something equally important: temporal scales. Just as with spatial scales, time scales define our experiences and perceptions in life as well as our view of the universe. Suppose there are two phenomena that are happening simultaneously, but one is happening much faster (say in seconds) than the other one (where changes happen over hours). A sketch of such a scenario is shown in Figure 6.2, with a fast oscillation superimposed on a slower oscillation. Well, depending on what time scales we are interested in, our description of the situation will be totally different. If we are interested in describing what happens over a few seconds, then for all practical purposes we can treat the slow process as static and unchanging, because that is how it will seem on the time scale of seconds over which the fast process happens. But, if we are interested in how the system changes over a time scale of hours, then it makes no sense to keep track of the details of

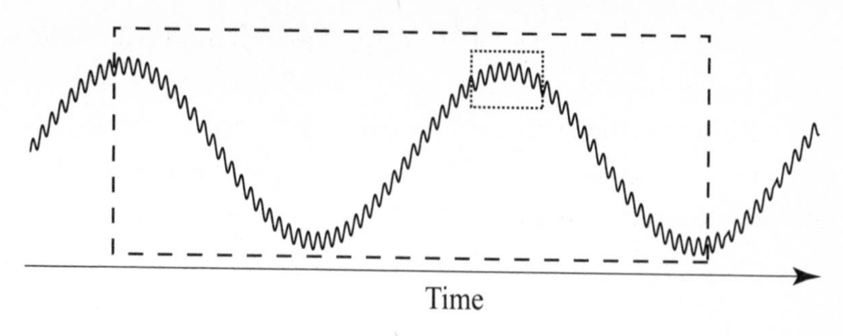

Figure 6.2 A fast oscillation is superimposed on a slower oscillation. If we are interested in what is happening at short time spans (as within the little box bounded by dotted lines), then we could focus only on the fast oscillation and neglect the slow undulation due to the slower oscillation. On the other hand, if we are interested in much longer time spans (as within the large box bounded by the dashed lines), then we could very well neglect the fast oscillations as just minor jitter on the larger and slower changes.

the fast process. We will only see a time-averaged effect of the fast process, and only the longer undulations of the slower process will be relevant. A simple example with a much less dramatic difference in time scales illustrates the point. Consider an oscillating table fan—we can easily observe the side-to-side oscillations, but since the processing time of our eyes is slow compared to the motion of the blades, they all blend together to look like a continuous disc. On the other hand, if we were to take a few pictures with a high-speed camera, we would see the motion of the blades in successive snapshots, but the side-to-side oscillation would seem frozen.

The average human lifetime defines the largest time scale for us. That time scale is implicit every time we talk about "forever" and "eternity," although really there are no such things in human experience, because "nothing lasts forever." But anything that seems to last a long time compared to our lifespans would appear forever to us; that is our closest brush with eternity—the illusion of eternity, that is. A diamond does not really last forever, but it will certainly last longer than your love, and your life, so the diamond companies may be forgiven for lying to you. The Grand Canyon has existed forever from our perspective, and for many of us, seeing it is an experience like no other because it gives a serene sense of eternity—so static, so immense, so unchanging—but, yes, only compared to our fleeting lifespans. But on geologic time scales, the canyon might seem to have been carved out overnight.

CHAPTER 7
STEREOTYPING STATISTICAL MECHANICS

If there is a loaded word in the English language, it has got to be "stereotype." It is hard to be objective about any usage of the word, so it may be heartening to realize that there are certain rules at work that gets to the heart of why stereotyping gets it wrong most of the time. The rules are those of statistical mechanics—the only scientific way to understand large multitudes of anything (or anybody, for that matter).

What could mobs, stereotypes, and polls have in common? Quite a lot, as it so happens. And I can think of few more entertaining ways to appreciate the power of the underlying principle than a few lazy weekends spent reading the *Foundation Trilogy*, Isaac Asimov's science fiction masterpiece that projects a highly imaginative vision of the distant future of humankind. The secret is in the thread that binds that series together—the idea that mathematics can be applied on a galactic scale to map out precise predictions about the common destiny of a quadrillion humans spread over millions of planets across the entire galaxy. The basic premise is that with such a large group to work with, even human behavior becomes highly predictable *en masse*, and human history becomes a precise, predictive science.

It is debatable whether predicting the future of humanity will ever become an exact science any more than that the entire galaxy would be populated by humans. But one never knows what fifty thousand years could do, considering how far we have come in just a couple centuries. Less ambitious versions of Asimov's general principle, however, have been with us for quite some time already, and they are with us to stay. Every time there is a poll of opinions, every time we predict the outcome of an election or a football game or forecast the volume of holiday retail sales, or anything of that sort, we are using similar ideas, which, these days, we call *the rules of statistics and statistical mechanics.*

The field of statistical mechanics is one of the most elegant and all-encompassing in the world of physics—and not just physics; it underlies pretty much all of science. Without it, we have little chance of describing the real world and the real universe in any realistic terms. Without it, physics would be stuck in the realm of idealization, with no room for the complexities and imperfections that are inevitably part of real-world systems. There is an inside joke among physicists, "let us assume a spherical cow," which is an admonition against over-simplification of problems in the quest for a manageable analysis, like for instance, if we are studying a cow, we might simplify the complexities of the bovine shape by "approximating" it with a more tractable geometric figure; never mind the details, we will add them in later once we have a basic understanding. Well, sometimes scientists can get carried away and take it too far, hence the "spherical" cow. Without statistical mechanics, much of physics would have remained forever locked up in an ivory tower of elaborate and beautiful laws and principles, of little practical use for describing the real world—as far off the mark as a spherical cow is from a real cow.

The reason is simple: Most fundamental laws of physics are elucidated in the context of a single object, like a ball dropping or a car moving or a bomb exploding, or a couple of objects, like the gravitational attraction between the earth and the moon. But in the real physical universe, we typically have to deal with a multitude of objects and sometimes unimaginably huge multitudes. Statistical mechanics provides the only practical and possible way to describe systems where there are many, as opposed to one or two, of anything—may it be electrons, atoms, stars in the galaxy, or galaxies in the universe. How important is that? Consider the fact that just one cubic meter of air (that is the air within the volume of a cube that is one meter, or about three feet, long on each edge) has about 10^{22} molecules (mostly oxygen and nitrogen) in it—that is 1 followed by 22 zeroes, or 10 billion trillion, a number that would make the US national debt (even when expressed in cents) look literally like a drop in the ocean. Even with all the laws of mechanics, from Newton to Einstein, from classical or quantum mechanics, there is no way, and no computer powerful enough, to keep track of all those zillions of molecules individually. So do we give it up as an impossible task? Absolutely not! This is where statistical mechanics comes to the rescue. It simply recognizes that we do not need to keep track of all the molecules individually to extract all the *relevant* information about such a volume of air. Taken as whole, the behavior is very predictable and can be described with a few very simple equations.

In fact, one of the earliest successful applications of statistical mechanics was precisely to explain the properties of gases (like air) starting from a molecular picture, and that theory is known as the *kinetic theory of gases*. By using the principles of statistical mechanics, kinetic theory reduces the description of most gases like air to just four quantities—its temperature, its pressure, the volume it occupies, and its weight—all of which can be easily measured; quite an improvement, huh? From 10 billion trillion to just four parameters to keep track of!

The principles of statistics and statistical mechanics are already prevalent all through the social sciences, and this is perhaps the most obvious and well-known application of ideas from physics and mathematics in the realm of social sciences and human behavior. Surveys and polls are pretty much based on the idea of average behavior being predictable, although individual behavior may be quite random and unpredictable. If you ask someone, "What do you think about the outcome of the next presidential election?" the answer is totally unpredictable, but if you ask a thousand people, a pattern will emerge, and you can say that on average 60 percent think the Democratic candidate will win versus 40 percent who think it will be the Republican. And the more people in the poll or survey, the more precise the results. If you somehow managed to poll all those who will vote, then it will be completely accurate. But that is usually not feasible—most people usually don't care for taking surveys even with some sugar-coated incentive. So polls always give a margin of error, which is a measure of the limitation placed by the limited sample of people polled. The larger the group polled, the smaller the error margin. The laws of statistics provide precise ways to compute both the averages and the error margins.

These days, polls and surveys are pretty standard, so all that I just said about average behavior is well established. Well, that was the easy and obvious part, like the two trivial and mundane questions they tend to ask at the beginning of a polygraph test! Now it's time get out of our comfort zone: Let's talk about the highly sensitive subject of stereotyping. Typically, people are averse to being stereotyped, but in reality everybody does it, whether consciously or subconsciously. The Oscar-winning movie, *Crash*, underscores exactly that fact—that everybody is guilty of stereotyping; even as people complain about being stereotyped, they are stereotyping others. The critical appreciation of a movie like that comes from a sense that it has a moral lesson for all of us: "Stereotyping is bad, and we are all complicit." But morality or a sense of right or wrong is not very useful for understanding stereotyping and can be quite misleading. Most of us are not lacking in morality and can distinguish right from wrong quite well. Yet we cannot help stereotyping. That's because regardless of the aura of rampant political incorrectness, we sense that at some level many stereotypes—not all—are based on some kernel of truth. While some stereotypes have absolutely no basis in reality, there are many others that simply hyperbolize facts, although often exaggerated and distorted beyond recognition. The fundamental rules of statistical mechanics will explain why and will even enable us to distinguish the ones that carry elements of truth from the ones that are complete fiction.

Nobody likes to be victims of negative stereotypes; everyone gets all up in arms with cries of "no fair" when negative stereotypes are applied to them. But you rarely ever hear anyone complain about being labeled with positive stereotypes. Well, we will need to deal with the bad ones if we accept the good ones! Whether we like it or not, there is some element of truth in many stereotypes. The best way to appreciate that is to consider some stereotype that is not personal for you, because none of us can be objective about any stereotype that applies to us, and we would be sure to vehemently disagree if it is negative.

But why do we stereotype? The answer goes deeper than most of us would care to acknowledge: Our minds are actually programmed to stereotype, because as humans we are compelled at every turn to draw conclusions from the limited data available to us. Stereotyping fundamentally relates to how we, as a species, learn, understand, and acquire knowledge. In science, it plays out all the time—when we know nothing about something, we start to look for patterns, we start to classify and compartmentalize; that is how knowledge begins. Before biology became a more exact science, biologists spent much of their time in the classification of species, going all the way back to Aristotle in ancient Greece. Even in the highly mathematical field of particle physics, when a whole menagerie of elementary particles started appearing in particle accelerators in the mid-twentieth century, the path toward a complete understanding started with classification of particles into different groups based on similar behavior and properties. One of the landmark breakthroughs, for example, was Murray Gell-Mann's organization of an important set of elementary particles called mesons and spin-half baryons into an octet structure that he whimsically called the *Eightfold Way*, a reference to a central idea in Buddhist philosophy.

Stereotyping emerges from that fundamental mental process of grouping and classifying applied to people. We have an instinctive need to understand the people and the world around us, and when it comes to analyzing people, most of us also tend to be supremely confident of our own abilities. But while our classifications of subatomic particles and even plants and animals might be factual and rigorous, our assessments of other people are anything but objective. Combine that with our inherent need to feel good about ourselves often at the expense of others, and the emergent stereotypes typically suffer significantly in accuracy. Yet we have no choice but to generalize at some level, because no matter how worldly we believe ourselves to be, our experiences are limited by time and space. It is a big world, and our lives are short, and the number of people we interact with is a very negligible fraction of humanity, even if we include all the people we see in the media. Nevertheless, our survival and proper functioning in this world predicates that we draw conclusions about the people and society around us and do so during the short time scales available to us; as a result, we tend to generalize from whatever experiences we have. For example, some guy spends three days in Paris, comes back to his small town, becomes an authority on all things French, and declares to credulous friends and family that the French are rude. Did he meet even a statistically significant number of them? Certainly not. Should we wonder whether his foreign mannerisms made him seem rude to the people he met, and whether their rudeness was simply a reaction? Quite possibly. We are all guilty of inaccurate generalizations of this sort quite often. While some of us may be more perceptive than others, every time we make any "informed" observations of any group of people connected by some perceived commonality—like race, nationality, demographics, community, neighborhood, etc.—based upon an inevitably limited set of interactions, we are stereotyping. But you know we all do that, and we will continue doing so!

From a statistical point of view, in the best-case scenario, stereotypes simply reflect the average behavior of a group of people as perceived by another group. Being human, of course, we are prone to exaggerate those behaviors—an obvious root of contention. But the real reason stereotypes get it wrong so often is because the emotion-driven and subjective human mind frequently chooses to forget the fundamental "no-no" of statistical mechanics: *that average group behavior cannot be strictly applied to individuals, and, vice versa, individual behavior absolutely cannot be generalized to a group.* They are two sides of the same bad-stereotyping coin, but they have different implications, as we will now see.

Sometimes we are familiar with the average behavior of a certain group of people who we might identify by their origin, sex, race, caste, education, profession, you name it, and we are aware of the stereotypes associated with them that are at least partially true for a large fraction of the group. Then we meet someone completely new who clearly belongs to that group and immediately we might assume that those stereotypes or average behaviors must be manifest in him or her. Although we might be right part of the time, it is also more than likely that we are wrong.

And equally as often, stereotypes are made exactly the opposite way: People had a certain experience, bad or good, with just one single representative of a certain group of people, and then immediately generalize the behavior reflected in that experience to the entire group. Examples: You got mugged in New York City once, so you label the entire city and its populace as dangerous. Or you meet one smart kid from your neighborhood college and assume every kid there is a genius. Unlike the previous scenario, this type of stereotyping almost never has any basis in truth. Because while average behavior is likely to be true to some degree for many members of a group, the nuances and eccentricities of one particular individual are far less likely to be shared by everyone in the group.

We can visualize this graphically. Let us consider a characteristic that won't offend anyone—say wealth. If statisticians from the Census Bureau summarize the annual household income of a particular borough, they are likely to use a *histogram* like the one shown in Figure 7.1. A histogram is simply a series of rectangles, in this case one for each income interval of $10,000, with the height of each rectangle proportional to the number of people in a particular income range. Notice we also draw a line that approximates the overall shape of the histogram. That line represents something extremely important in the study of statistics. The histogram here

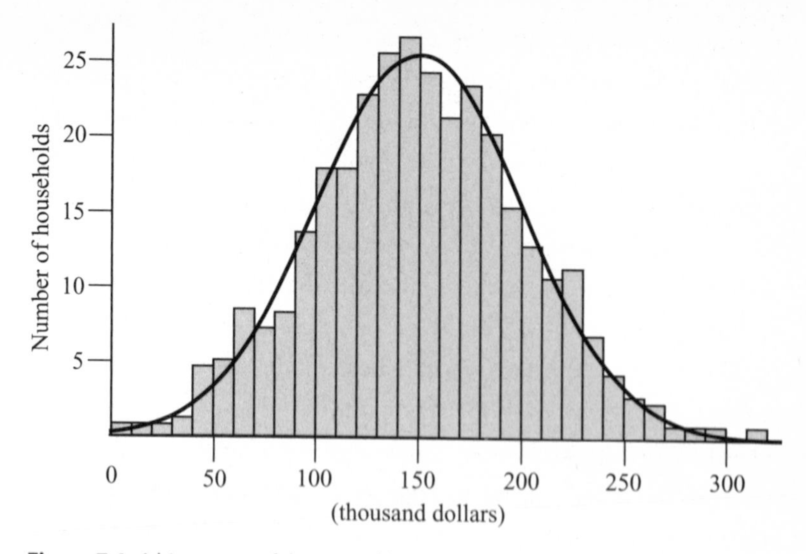

Figure 7.1 A histogram of the annual household income of a borough is shown. The height of each rectangle corresponds to the number of households in each income bracket of $10K increments. The smooth curve that approximates the overall shape of the histogram has a typical bell-shape and represents what is known as a normal distribution in statistics.

is obtained by sorting the incomes in bins of $10K, but the curve gives a mathematical formula to compute the approximate number of people at *any* income value. For example, while the histogram can only tell how many people make between $120K and $130K, the curve can give an accurate estimate of exactly how many people are making, say, $126K. Curves like these have the intuitive name, *distribution functions*, as in this case, it happens to represent the *distribution* of household incomes in that borough. The particular *bell-shaped* distribution that we see here in the figure is the most common, as well as the most important, one in the field of statistics, and you can tell how much so by its name—*normal* distribution. One of the most fundamental results in probability and statistics, the *central limit theorem,* states that if you take the average of a large but fixed number of random variables (like the average of the numbers that appear face-up in tossing a thousand dice simultaneously), and keep on doing so multiple times (keep tossing the dice hundreds of times—the average would vary with each throw, since this is quite random), and then you plot how many times you get a particular average value, you will *always* end up with a normal distribution like the one shown here. It's a very powerful statement because it predicts a very precise mathematical distribution for a set of *random* and arbitrary variables! This gives a glimpse of why normal distributions are so important in statistical modeling, because finding patterns in randomness is the very name of the game—with obvious relevance for a bulk of important things in science and in life!

Now if you look at the distribution function in the histogram in Figure 7.1, you will notice that the average income in this borough seems to be about $150K. Suppose you already happen to know that. Then if you meet someone, you would certainly do a bit of stereotyping that the person is well off in the upper middle class but perhaps not a multimillionaire. And, since most of the people seem to have incomes between $100K and $200K, there is a good chance that you would be right. This is an example of the first kind of stereotyping. Since you know the average, and most of the distribution curve lies close to the average, you know that your guess is more likely than not going to be close to that average. However, you can also see in Figure 7.1 that there are quite a few families making less than $60K and a fair number making above $250K, which means that your stereotyping of the person that you just met could also very well be way off the mark.

On the other hand, suppose you know nothing about this borough, and the very first person you meet from there is someone who was laid off two years ago and had just a small income of about $15K last year from doing some odd jobs. Now, if you jump to the conclusion that this borough is a poor one, you would be

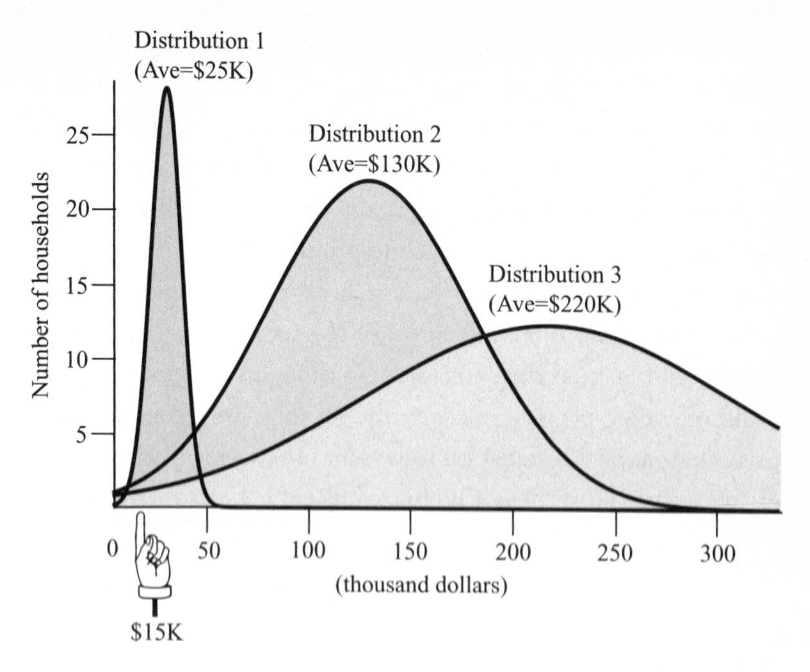

Figure 7.2 If the first person we meet from a borough had an annual household income of $15K this year, from that we cannot conclude anything about the income distribution of the borough. Three (of the infinite possible) distributions are shown here, all containing $15K within their ranges, but each one has a very different average—and anyone or none of them could be right!

very wrong. This is the second kind of stereotyping. In this case, with absolutely no knowledge of the average or of the wealth distribution in the borough, and just one sample point, that point could belong to any kind of distribution, because we could draw infinite possible distribution functions around it; for example, in Figure 7.2, we see three possible ones with $15K within their ranges. If the actual one is Distribution 1, your guess might have been correct because the average income is $25K, but if it were Distribution 2, or even more so, Distribution 3, then $15K is an outlier very far from the average income in the borough, and you would be completely wrong in your presumption!

Basic rules of statistical mechanics therefore confirm and explain why, even in the best-case scenario, we are quite likely to be wrong when we apply stereotypes presumptuously—but in some cases more than others. With this bit of insight on hand, can any of us commit to never stereotype again—never label people, places, or situations, with our own preconceptions before getting to know them? Probably not. We just have to accept that people will never stop stereotyping; it is built into us. The socially smart among us simply tend to keep their stereotyping thoughts to themselves and avoid offending people unnecessarily. Yet there are a couple of very useful lessons in this statistical viewpoint. The average-behavior sort of stereotype can serve as a useful reference, if we can make a conscious effort to withhold final judgment on specific individuals until we actually get to know them, because as in nature, the behavior of each individual can be completely unpredictable and significantly different from the average. But the more important message is that *it is never a good idea to generalize from a single experience or a sample space of one—neither in doing science or with people*!

Let us move on to something much less touchy. Mob behavior is also often predictable because it involves large groups of people. Well, this is what happens: Throw in a whole bunch of different types of people in a common situation—like rooting for the same team or having the same grievance—that, at that moment, is the one and only thing that they all share. All other aspects of their personalities and thoughts get averaged out because they are certainly all different for everyone. As a result, averaged over the group, the one thing they all share gets amplified through their interactions with one another, and that can sometimes lead to serious mayhem in the form of riots. Soccer (football outside the United States) hooligans build on their fanatical shared support for a particular team; lynch mobs amplify their common hatred for some individual or group.

Statistical averaging gives a sense of how mob behavior emerges in a group, but there is another very subtle principle of physics at work in how exactly a shared characteristic can get amplified by interaction, but we will have to wait until Chapter 14 to see that.

To come full circle back to Asimov's vision of the future of humanity—aren't we already partial believers when we say, "History repeats itself?" After all, anything that repeats is predictable. Because in the end, on average we humans are much more alike than we are different, and on the scale of large numbers of people and over timespans of decades and generations, our behavior is more predictable than we would like to admit. On average, we know how a human life plays out in any part of the world: spend early years getting educated and then the all-consuming search for a mate, get married, get a job, build a career, buy a house, have kids, kids grow up, you grow old, develop various levels of health problems, worry about your finances and retirement plans, perhaps move around from place to place, job to job a few times, and then you retire, and so on. There are many variations of these, some drastically different from others, but on average this is what plays out for most people in most parts of the civilized world.

While history does not always repeat itself exactly, the patterns of human behavior *en masse* certainly remain surprisingly uniform across time as well as space, so that in a broad sense certain aspects of history do seem to follow recurrent patterns. But such recurrences have long enough periodicity—perhaps several generations—so that it is easy to overlook them altogether since our collective memory is short and seems to reset after a generation or two. The recent financial crisis caught most of us by surprise; well, if we read our history, it is just the last one in a long line of similar crises brought on by eerily similar causes—particularly uncontrolled greed. Or take wars, for instance. Nobody wants wars; they can fool us all they want with the propaganda about the glory of war, but nobody wants to get killed for glory, and yet there are wars all the time, because the same dynamics of conflict play out over and over again with groups and nations, large and small: the aggressor who wants to take control of something (material or ideological) that the other has or refuses to give up, and the defender who is compelled to resist. Typically, both believe that God is on their side. Likewise, all the empires ever built were generally convinced at their height of glory that they will last forever, but they all vanished, and they always will. The list goes on. History might never become a predictive science, but patterns of collective human behavior are predictable enough that we ignore history at our own peril.

CHAPTER 8
FLUCTUATION WILL LEAD TO DISSIPATION

With instant connectivity and instant just-about-everything, there seems to be a conspiracy of distractions to keep us from getting anything done these days. In escalating the battle against this onslaught of distractions, success can hinge on appreciating Einstein's insights on how fluctuations (nature's version of distractions) directly correlate with dissipation of useful energy.

Nobody is perfect. Perhaps you are the one exception, but I will take my chances. We learn early on that perfection is elusive and certainly so among people. Inefficiency and imperfections are rife throughout life, slowing us down, distracting us and sometimes leading us to failure—and quite often we have only ourselves to blame. But have you ever wondered why? Why is it so hard to stay the course, to face up to every task and challenge in life with complete efficiency, and to give it all we have got? Why can't we naturally stay focused 100 percent on our goals all the time and instinctively optimize our efforts? Standard wisdom would have us believe "to err is to be human," so does it really have something to do with being human?

The answer is a most definite no! The cause of much of human failure and inefficiency actually lies within the much broader scope of an all-encompassing law of nature called the *fluctuation-dissipation theorem*—a profound law that directly ties the loss or dissipation of energy from a system with the inherent fluctuations of its state. This law has deep implications in our lives, as well as in a wide range of subjects from the basic sciences to finance and economics.

If you have ever felt frustrated with yourself or others about not getting things done efficiently or quickly enough; if you spend a lot time trying to catch up at the last minute; if your goals always seem farther and harder to reach than you initially thought; if you are uncomfortably aware that you waste too much time and energy on pointless activities like watching TV or Googling; if you never can seem to get through the day without a million interruptions; if you . . . well, you get the idea, then you are an unwitting victim of the fluctuation-dissipation theorem. This list of "ifs" is a long one because it is the list of our common frustrations, and we all share most of them to some degree. The fluctuation-dissipation theorem hobbles us all. But the good news is that the theorem also has within its very statement the secret to reducing its impact on our lives. But the better news might be that as a basic law of physics, it is ubiquitous and impossible to circumvent completely—so we might as well stop stressing about the little annoyances of life and accept that inefficiency and imperfections are built into the universe we inhabit; there is no use beating ourselves up about it any more than decrying the limitations that prevent us from living forever. In the end, the fluctuation-dissipation theorem gives a scientific perspective on how and why we all inevitably err in our lives, one time or another.

So, what is this theorem all about, and how does it affect us? The theorem is actually one of the most important laws in the field of statistical mechanics that we discussed in Chapter 7. Although it was formulated and perfected by many scientists over the years, its true origin lies in one of Einstein's four ground-breaking papers in his miracle year of 1905. The contents of three of those papers are widely known and have even seeped into popular culture: one gave us relativity; another created the most famous formula in science, $E=mc^2$; and the third one truly laid the foundations of quantum mechanics by using light quanta to explain the photoelectric effect.[10] But not many outside of physics have heard of the fourth paper. Yet that fourth paper is his most cited work among them, and, in terms of practical usage and real-life applications, this was perhaps Einstein's most important paper with applications in everything from physics, chemistry, and biology to finance and economics. This was his paper on *Brownian motion.*

Robert Brown was a Scottish botanist who lived in the nineteenth century long before Einstein. In 1827, the observant Brown noticed that if he scattered some pollen grains on the surface of a bowl of water, the grains moved about in random and erratic jittery motion, even when the bowl and water are absolutely still with no currents of air to disturb the water surface in the slightest. In case you are thinking it has something to do with pollen grains being "alive," it turns out that the same thing happens with dust particles as well. If you have access to a microscope, you can try and see this for yourself. The big mystery was, since the water itself was perfectly still, what was causing this motion? There were many speculations, of course, but the key idea needed was an ancient one overlooked by all because very few really believed in it. And that idea, which was first proposed by Democritus of ancient Greece about 2,500 years ago, was that all matter was composed of

[10] When light (hence "photo") shines on certain metals, it creates an electric current, because the energy in the light knocks out electrons from the metal. This is used, for example, in automatic doors that open when our presence blocks an invisible beam of light and interrupts the photoelectric current.

tiny particles, which he called *atoms*. Incredible as it may seem today, even at the beginning of the twentieth century when the world was well on its way into the modern technological age, scientists still had not come to terms with the existence of atoms and molecules, which even kids learn about in elementary school today. In fact, Einstein's solution for Brownian motion is considered to be the final bit of hard evidence that pushed the idea of atoms from being just a good hypothesis to the realm of solid, indisputable scientific fact.

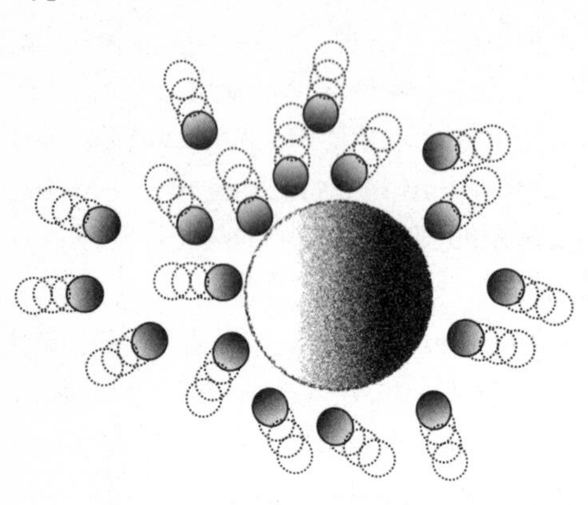

Figure 8.1 Pollen grain bombarded by tiny molecules of water push it one way or another randomly, depending on which side is more impacted at any instant.

In retrospect, the explanation is quite simple. Water consists of countless tiny molecules moving in a random fashion. Each water molecule is comprised of *two* hydrogen (H) atoms and *one* oxygen (O) atom bound together—hence the often-used alias, H_2O—but that is not important for the explanation. What matters is that the water molecules are moving around at random and that they are much smaller than the pollen grains. Each pollen grain, although small, is a giant compared to the much tinier molecules of water, and each floating grain is constantly bombarded on all sides by these tiny water molecules. Since the collisions are random, at any given time the net effect of all the collisions is that each pollen grain moves in one direction then in another depending on which side happens to have more water molecules crashing into the grain at that particular instant. It is as if the pollen grains had dropped in on a never-ending game of molecular bumper cars. Consequently, every grain zigzags about at random, bumped around by the random collisions with the water molecules. This is just what Robert Brown observed.

But Einstein did not stop at just giving a qualitative explanation like this. He actually derived a precise mathematical formula for how far, on average, a pollen grain could be expected to have moved away from its initial position after any given period of time. As it turns out, that equation he derived was a special case of what was later generalized into the *fluctuation-dissipation theorem*. Einstein showed in his characteristically simple, but incisive, way that the rate of diffusion or drifting away from the initial position of a grain of pollen is directly related to the amount of energy lost from the system, manifest as thermal or heat energy. His theory pointed out that the random impacts that cause fluctuations are also the ones responsible for the frictional forces that lead to loss of energy of anything moving on the surface of the fluid. The connection is easy to see if we imagine trying to pull the pollen grain along the water surface, then the colliding water molecules will provide drag acting to slow down the motion, leading to dissipation or loss of energy that the grain would have due to its motion. Imagine walking through a crowd that is just milling around, you would need more time and more energy to

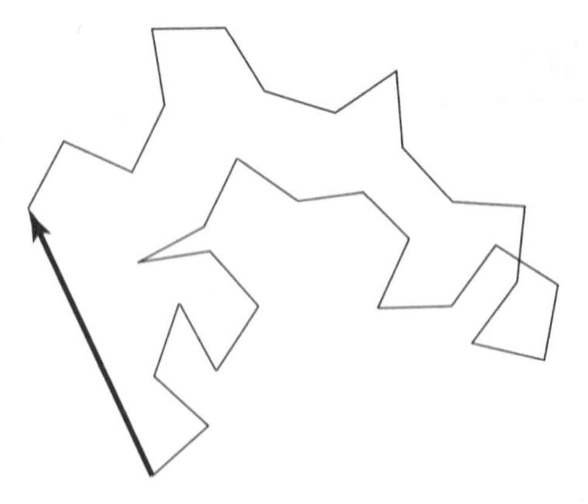

Figure 8.2 In Brownian motion, the net distance (bold arrow) covered is much less than the total distance "walked" randomly.

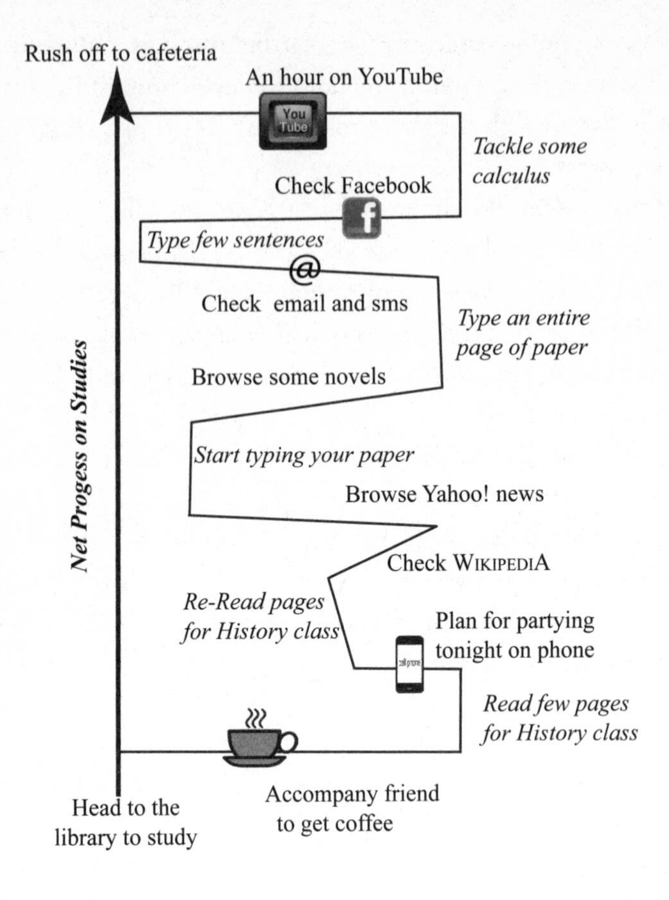

Figure 8.3 Fluctuations (distractions) significantly reduce the net productive work (indicated by the bold arrow) done for the total amount of time and effort spent. Progress upward indicates useful work (indicated in italics); sideways indicates all the distractions.

cover the same distance than you would if it were completely empty. At the same time, the crowd is also responsible for you zigzagging among the people as you walk across the room.

The fluctuation-dissipation theorem is strongly with us these days, having been given a significant boost by technology. Look at the illustration in Figure 8.3, and note its qualitative similarity with Figure 8.2. In the next couple of pages, we will accompany a typical college student along this tortuous path laid out in Figure 8.3 and get an appreciation of just how the fluctuation-dissipation theorem defines his progress as he tries to catch up on his reading and homework assignments on a Saturday afternoon. He sets out toward the campus library, full of purpose and energy. Halfway there, he runs into a friend, starts chatting, and is easily persuaded to walk with her to grab some coffee. Twenty minutes later, coffee in hand, he finally bids adieu and continues on toward the library, makes it there with no further interruption, finds a secluded corner, opens his books, turns on his laptop, and settles down for a productive afternoon. Then, of course, the phone vibrates, "There's a party tonight, you wanna go?" He virtuously replies, "I would love to, but dude I have to study, but can you call back after dinner?" The negotiations take up another ten minutes. Now where was he? Ah, page 4, but now he has completely forgotten what he had read, so he starts again from page 1. He settles down and is finally making some progress with the reading, but then he encounters something he does not understand, so he goes online to check it out on Wikipedia. While opening his browser, something catches his eye on Yahoo! news, so after Wiki, he starts reading that, and continues on to read several of the mindless comments at the end of the news article, and then fluidly moves on to some other pieces of gossip about assorted celebrities and reality stars, and before he knows it, another half hour has gone by. Ah, now he is truly sick of reading in general, so he closes his book and decides that he might as well start writing the paper that is due Monday, opens up Microsoft Word, and begins typing. But he has not read the material through, so no wonder he does not get very far with the writing; he must have writer's block, or so he tells himself. Perhaps walking around a bit will clear the thoughts, and besides, that large coffee is having an effect, he needs to use the restroom anyway. He eventually makes it back to the table and for the next fifteen minutes hammers away furiously at the keyboard and produces an entire page—double spaced. He sits back for a second, and then suddenly realizes that he has not checked his email today, so he logs on, and there are a couple of funny ones from his sister, so he has a good laugh, quietly though—it is the library after all. He shoots off an appreciative reply. Well, while he's at it, he might as well check his text messages—reading, deleting, and replying takes another fifteen minutes. Back to the paper—he types a few sentences while still thinking about the messages he just read. Pretty soon he finds himself checking out his Facebook news feed. After half an hour of that, he begins to feel some guilt creeping up and tells himself, now it is time to get serious—but he has run out of inspiration for the paper. Why not try something completely different? He reluctantly opens his calculus textbook and stares at the problems for an intense five minutes, but he has no idea where to start, and it is too late to go to the tutoring center. Perhaps he can do the first problem; he tries and thinks he has figured out the first few steps, but it is still too hard. To hell with it, why not take a break; he opens up YouTube, puts on headphones, and just forgets time. Before he knows it, an hour has gone by, and he realizes with alarm that the cafeteria is about to close and if he does not hurry, he might miss dinner. He hurriedly packs everything up and walks toward the cafeteria feeling exhausted and frustrated that after four hours in the library, he only got about half an hour's worth of work done. All that energy he started with, all dissipated and very little progress to show for it. We can follow the student's trajectory for the afternoon in the sketch in Figure 8.3—isn't it very similar in spirit to the trajectory of a particle undergoing Brownian motion in Figure 8.2?

But of course, you know that this scenario is not unique to college students; we are all victims of endless distractions all the time, exacerbated these days by instant access to information, music, videos, friends, and relatives via cell phones, Twitter, texting, and email. Such distractions or fluctuations are the primary cause of much of human inefficiency and frustrations with consequent lack of progress, and therefore they are the root cause of most of our failures. But that's not all—the fluctuation-dissipation theorem has other implications in our lives.

A hesitating mind is a fluctuating mind. Consider one of those people who can never make up her mind. She stands in front of the vending machine for ten minutes, debating between Coke and Sprite, and a significant fraction of her life has been spent in choosing between chocolate and vanilla. And for things that really matter, it takes her so remarkably long to make any decision that the time window for making the choice keeps passing her by, like for instance, "Should I talk to that man or should I not?" She is a victim of the vacillations of the mind, fluctuations of the will . . . and the result is a life strewn with missed opportunities and the gradual loss of will and energy to succeed—the fluctuation-dissipation theorem at work again.

Or consider the guy with the roving eye; he is married, but he can't keep his physical senses focused on his loving wife. At first, he justified it by saying, "It's alright to check the menu, as long as I dine at home," but pretty soon he finds himself ordering the full course with dessert and port wine to follow . . . he's a fluctuator—a possessor of unfocused attachments. And we all know what that usually leads to—dissipation of energy in so many ways: mental energy in the schemes to cover up the affair; and when discovered, vocal energy in shouting matches with the jealous and furious wife; financial energy in the form of alimony and lawyer's bills in divorce court; more mental energy in the anguish of lost love and separation pangs from children; binding energy in the form of lost attachments and love. Fluctuation again leads to dissipation.

The fluctuation-dissipation theorem is actually the root of some of the most pertinent wisdom about how to achieve general well-being in our lives. For one, it is the ultimate justification for anger management, if we ever needed one. Anger is a severe fluctuation of our mental state leading to noisy, unpredictable, destructive behavior, and we all know that at the end of it, a lot of energy is dissipated all to very little purpose and usually with detrimental effects. Conventional wisdom about the right approach to life, our quest for tranquility and calm, the belief that slow and steady wins the race, our collective distaste for erratic behavior and aberrations, our appreciation of stability—all of these arise, in a sense, from the fluctuation-dissipation theorem. A mechanical analogy highlights just how: When we drive a car, we get the best gas mileage by maintaining a steady speed; that is why highway mileage is always better than city mileage, because in the city the speed fluctuates all the time due to heavy traffic, stop signs, and red lights. It is the same way with people—if we maintain a steady course in life in any endeavor, then in the long run we make much more efficient use of our time and energy. Carrying the mechanical analogy one step further, every mechanic and engineer knows that reducing noise and fluctuations is at the heart of getting the best performance out of a machine; likewise reducing irregular behavior and distractions is the secret to getting the most out of life, both mentally and physically. Any doctor will tell you the importance of regular habits—sleeping, eating, and exercising regularly are among the best things we could do for a healthy life—our bodies hate fluctuations. Stress often results from fluctuations of our rhythms; modern life deprives us of the daily eight hours of sleep we need and often plays havoc with the regularity of our meals and exercise schedule. The result is stress, stress, and more stress; a lot of physical, mental, and functional efficiency lost; and oftentimes serious breakdown in the form of various ailments and diseases. So really, at the root of it all, this is not just about biology, medical wisdom, or the human body; regular and steady behavior is mandated by a fundamental law of physics: the fluctuation-dissipation theorem.

If you are in good health and are indifferent to threats to your mental and physical well-being, you still can't ignore the effects of the physics of the fluctuation-dissipation theorem in your life, because it also directly affects something no one can "afford" to ignore: money! Brownian motion and the fluctuation-dissipation theorem hold the key to how the technical aspects of high finance work. You might have heard of "rocket scientists" in popular culture; well, these rocket scientists have nothing to do with rockets, not even bottle rockets. They usually refer to physicists and mathematicians hired by Wall Street to analyze and predict pricing of stocks, bonds, options, and derivatives. In everything they do, they heavily rely upon something called *stochastic calculus*, and stochastic calculus is built upon a fundamental property of Brownian motion that Einstein derived in his seminal paper of 1905. In fact, all genuinely mathematical predictions of the stock market and the pricing of options and derivatives are based upon stochastic calculus in some way, and the mathematics of Brownian motion and the fluctuation-dissipation theorem plays a crucial role. The underpinning of this calculus is based upon the fact that random processes like Brownian motion do not depend upon time linearly like in ordinary time evolution. Let me explain. Say you are walking decisively along a straight path and cover about 1 meter in each stride, and you take one step every second, then in 50 seconds you would cover 50 meters, and in 100 seconds, 100 meters. So the distance you cover is directly proportional, or more specifically linearly proportional, to the time that passes; if we plot the distance covered versus time, we would get a straight line as we can see in Figure 8.4. Well, not so with Brownian motion; in fact, we call such motion a *random walk*, because at each step you can move in any direction, like in Figure 8.2. As you can imagine, such motion would not be very efficient, it is like a really drunk guy staggering around looking for the way back to his house from the local bar. In fact, for Brownian motion, it turns out that there is a *square-root relation*: so in 64 seconds you could cover a net distance of 8 meters on average from your starting point, in 100 steps only 10 meters, and so on,[11] and therefore, as shown in Figure 8.4, Brownian motion covers significantly less distance over the same period of time. Unlike ordinary calculus (used to predict trajectories of ordinary motion like that of a thrown baseball), stochastic calculus inherently assumes this square-root dependence on time.

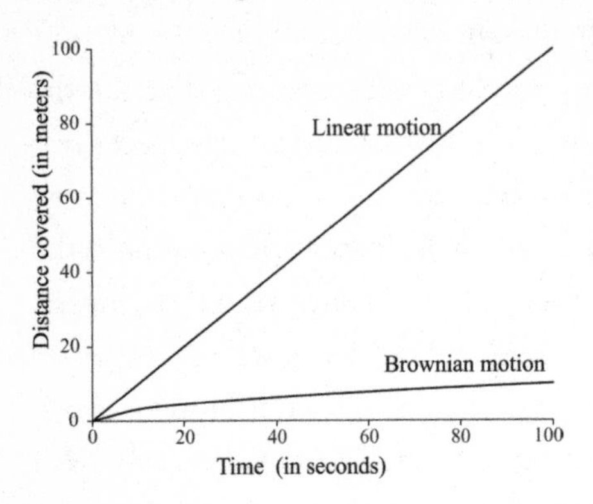

Figure 8.4 The large difference between ordinary linear motion and Brownian motion can be seen here. If person takes one step every second and walks in a straight line, he would cover 100 meters in 100 seconds, so the distance covered is linearly proportional to the time. But in Brownian motion, he does a random walk zigzagging about, and on average the distance covered would be only about 10 meters (which is the square root of 100) in 100 seconds.

How is this relevant in finance? You see, the price of stocks and bonds and other financial products behave very much in the

[11] The square root is the opposite or inverse of taking a square of a number, which means multiplying a number by itself. Thus, the square of 8 is 8 × 8 = 64, so the square root of 64 is 8. Likewise, 10 × 10 = 100, so the square root of 100 is 10, and so on.

Brownian way; it is quite random on short time scales, although patterns might emerge in the long run. It is random because of the same reason that the motion of Robert Brown's pollen grains was random—because there are too many factors influencing the price to keep track of, just as the pollen grains are impacted by too many water molecules to keep track of. Therefore, the movement of the price and value of those financial products and instruments can be most accurately tracked and predicted by treating them literally like Brownian motion, with the price fluctuating up or down in a random manner, and of course the fluctuation-dissipation theorem also plays a role. The beauty of stochastic calculus and the physics of the fluctuation-dissipation theorem is that you can still make somewhat accurate predictions even for such seemingly random processes. Otherwise, these rocket scientists would not be so highly paid. But it is not always accurate; if it were, the stock market would be completely predictable, and nobody would lose money. Well, one obvious catch is that the very process of prediction and the resulting actions themselves become part of the factors influencing the market!

It is worth noting that in a wide variety of situations where there is random behavior, the physics of Brownian motion and the fluctuation-dissipation theorem can be directly used. Of course, both in nature and in human society, there is a significant amount of randomness inherent. Therefore, beyond physics, these ideas are highly relevant in many other fields, such as biology and sociology (to predict the evolution of populations, for example) and of course in finance and economics, as we just saw.

Beyond practical applications and relevance in our habits and behavior, the fluctuation-dissipation theorem actually provides a measure of the intrinsic quality that makes us humans a unique species and that has led to the advances in knowledge, science, and technology that separates us from the other species on the planet. What really differentiates humans from other creatures is not just our superior intelligence but our capability for focused thought. Intelligence among people is hard to measure and even harder to compare—how can we compare the intelligence of a desert-dwelling Aborigine that allows him to survive in his extreme environment with that of a rocket scientist good at making money on Wall Street? The intelligence of one is useless in the other's environment. But there is one surefire way of gauging mental capacity, and that is the capacity to focus. A chimpanzee, although supposed to be one of the few species that is self-aware (at least it passes the mirror test—it can recognize its reflection as its own image), cannot focus on anything for very long, and as we go further down the evolutionary ladder, you will see less of a capacity for focusing on a thought (that is, if there is any capacity for thought at all) or even on an object, except maybe by instinct of a predator on a prey.

At the highest extreme, we have human geniuses; it is well known that Einstein and Newton, considered the two greatest physicists of all times, had incredible capacities to focus on an idea or a problem with superhuman intensity for long periods of time. It is truly very difficult to do that. Just try it—take a single difficult thought or a problem and just try to think of it and nothing else for ten minutes or even just one minute, and I mean absolutely nothing else—you will quickly see how challenging that can be.

But before you go off and convince yourself that you have genius-level focusing capabilities, let me clarify. Well, you truly may be a genius, but I would just like to make sure you do not mistake brooding for focus. You can brood continuously for days about how much you hate your co-worker or how someone ticked you off, and you could not do anything about it, or you can keep on turning round and round in your mind some incident that embarrassed you so much the other day. No, that is not what I mean by the capacity to focus. If you really want to test your capacity for focusing, take a hard concept that you have difficulty understanding—a physics textbook will provide abundant material—and then try to understand that without letting your mind wander. That is what I mean by the capacity to focus such as what Einstein and Newton had to such a high degree. The truth is that for most of us, most of the time our minds are continuously distracted by random thoughts and various sensory inputs—our thoughts fluctuate—and the result? Well, you guessed it, dissipated energy and effort and reduced productivity. Actually, there is indeed an ancient and proven technique for improving focus. It is called meditation. The key idea in meditation is that you focus on one thing—an object, an idea, a light source whatever—and if you succeed in doing this for long enough time, you might succeed in eliminating all those fluctuations in your mind, and I suspect that is what they call "a state of bliss," where the mind transcends its typical Brownian motion of random thoughts. Ancient Indians considered this the source of true human power and spirituality, and in recent decades it has been catching on in the rest of the world with great positive impact.

Apart from its many practical applications, we can now see that the fluctuation-dissipation theorem truly contains the root cause of most of our failures in life—why things never go as smoothly as planned! Just as in nature a system loses energy and momentum due to various fluctuations, so do we in our lives often lose sight of our goals and yield to all the distractions and temptations of life and end up losing steam and never getting there. Fluctuations are an inherent part of the reality of this universe; fluctuations fundamentally differentiate the ideal from the real, in both nature and in life. All we can do is try to reduce it, but we can never eliminate it completely. "Nothing is perfect"—the truth of that really goes much deeper than we usually assume, because randomness and fluctuations are inevitable in complex systems, and human life and society are about as complex as it gets. As you face your daily trials in the court of life, you can upgrade from the lame frail-human defense, "To err is to be human," to a law-of-the-universe blanket excuse for all your failings, "To err is to obey nature," or more precisely, nature as dictated by the fluctuation-dissipation theorem.

CHAPTER 9
THE WAVE MECHANICS OF RELATIONSHIPS

Relationships are perhaps the most confounding of life experiences. Practice seems to make less perfect, and endless analysis distills more confusion! So, here, we bring a fresh, new perspective to this universal issue by applying the principles that govern the motion of waves everywhere—in the ocean, in light, and in sound. By systematic mapping of personalities to unique waveforms, we define an algorithm for compatibility of people in relationships based on constructive or destructive interference among the waveforms. If the chapter seems a bit complex, well, no one ever said understanding relationships is easy!

People have been trying to understand human relationships forever, analyzing why we love or hate, like or dislike, tolerate or despise the different people in our lives, who we get along with and who we don't, and why. Art, poetry, literature, psychology, sociology, and pretty much all the branches of the humanities, have been brought to play in the never-ending analysis, and lately even some biology as in effects of pheromones and genetic predispositions. Magazines, talk shows, and sitcoms all revolve around relationships. Yet in the end, we remain as confused and bemused by relationships as ever. If that were not the case, most of those magazines, talk shows, and sitcoms would be out of business. Just look around—the divorces, the breakups, the make-ups—and always, in the end, people are left debating about what really happened, where it went wrong, or if it did go right, how so? Social conversations, text messages, and cell phone minutes are largely dedicated to discussing and dissecting relationships from every possible angle. With this much human energy and effort devoted to deciphering it, you would think we would all be relationship experts by now, yet we still do not seem to have a clue.

Why? Because our views about relationships are never objective. They are invariably derived from the biased perspectives of our own personal history, or the subjective opinions of self-proclaimed gurus and experts pushing their own agenda, or the partisan advice of friends and relatives just trying to make us feel good. What we need is a fresh approach—something that can cut through all the subjectivity and the ambiguity. In short, we need to unleash the laws of physics on human relationships—to give us the edge of a hard scientific method that is generally missing in our usual attempts to decipher this most confusing of human experiences. Some of the concepts we will encounter along the way might seem a tad difficult but with good reason; after all, we are trying to explain one of the most complicated things in life, so you know you can't expect it to be easy!

Relationships are hard. It is hard enough to get one started—just to find that someone to have a relationship with. Then once we are through that bit, the realization sets in soon afterward that perhaps getting started was the easy part, compared with what we are up against now—to keep it going. When that realization hit me for the first time, I was still under the Beatles-vocalized delusion that "love is all you need." I was just a young student, and I did not know any better; earlier relationships never seemed very serious and never lasted long enough, for that matter, so I thought I had it all figured out. Right! Until a real relationship happened, and I could tell the difference quickly enough, like soldiers feel when real bullets start whizzing by after training's over. For one thing, even when things inevitably got rough—a time in which I would usually find myself single again trying to move on—now there was this strange compulsion to stay together come what may. I also discovered, much to my surprise, that love was not enough—not by a long shot! I finally had to face the facts: I actually knew nothing about relationships.

Amidst bouts of joy and misery, interspersed with pointless arguments that everyone goes through in a failing relationship, I desperately looked for solutions. But as we all find out sooner or later, standard logic does not seem to apply to these situations. Then, one morning, as I sat half asleep in my quantum mechanics class, brooding over the crisis, I found some answers in the most unlikely of places. I realized that, hidden underneath all the irrationalities of relationships, there are actually some simple rules after all, and I had been looking for them in the wrong places. The key to understanding how relationships work or fail lay not in love, but right there among those *waves* the professor was drawing on the board.

Waves are everywhere, although we might not always realize it. In the quantum view of the world, everything in the universe is an expression of waves of probability density; sound and music are all about waves, light is wavelike most of the time, and if we go further down in the scale of things, from atoms, to nuclei, to quarks, to the realm of string theory, we encounter waves of strings and membranes that supposedly underlie all of reality. Waves are indeed ubiquitous in the universe; therefore, it makes perfect sense that waves could also explain much of human behavior, however irrational and quirky it can be. But before we analyze the *wave mechanics of human relationships*, we need to review a few things about waves. It is all quite simple and will just take a few paragraphs and a couple of pictures here and there to gather all we need to know.

All of us actually know more about waves than we might give ourselves credit for, and it is just a few small steps from the everyday notions of waves toward a more precise definition as used in the sciences. There are waves in the oceans and on the seashore; when you toss a pebble in the lake, you see ripples spreading out—that's a wave; any sound you hear is due to wave motion of air masses striking your eardrums; you might have seen and done the wave in a football stadium; and you have probably heard that even light is really an electromagnetic wave.[12] But among all these diverse phenomena, what is the common element that characterizes them all as waves? It is that they

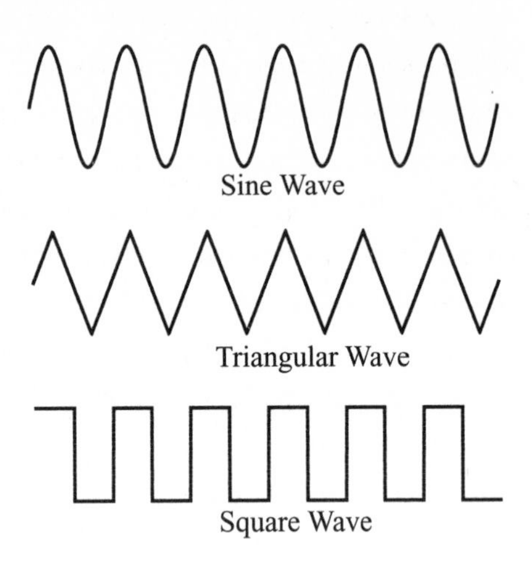

Figure 9.1 Some examples of different kinds of waves are shown. The common feature of all waves is that a certain pattern repeats itself.

[12] *Electromagnetism* refers to the related phenomena of electricity and magnetism. With both phenomena there is a region of influence; for example, a magnet can attract iron objects in its neighborhood. That region of influence around a magnet or an electrically charged object contains an invisible electromagnetic field by which such influence is exerted. Any disturbance in that field propagates like ripples in water and is called an *electromagnetic wave*. Light originates in the disturbance of the electromagnetic fields associated with moving electrons inside atoms, as we saw in Chapter 1.

all display some *cyclical or repetitive pattern or behavior* over space and time. Thus, the ripple in the lake displays repeating patterns of crests and troughs; people doing the wave in the stadium stand up, wave their arms and sit down in turns in a pattern that goes around the stadium; sound is caused by alternate compression and decompression of air.

It is easy to draw wave patterns on paper, and they come in all forms and shapes—from the very simple to the extremely complicated; we can see a few simple ones in Figure 9.1. The simplest and, therefore, the most important wave of all is called a *sine* wave, and that comes from the name of the trigonometric function that describes it mathematically. In fact, if you were to ask kids to draw waves, they would often draw something that resembles a sine wave; it is the natural shape we all associate with waves. And because it is so easy to visualize, we will use a sine wave to illustrate the general features of waves.

Since waves are cyclical, the fundamental feature of a wave is the distance over which it repeats itself—that distance is called the *wavelength*—and for a sine wave, that happens to be just the separation between the successive crests, shown in Figure 9.2. Waves can be strong or weak—is it a tsunami or just a ripple? The strength of a wave is measured by its *amplitude*, which is the difference between the height of the crest (the maximum) and the bottom of the trough (the minimum); the larger the crest-to-trough difference, the stronger would be the wave—ask any big-wave

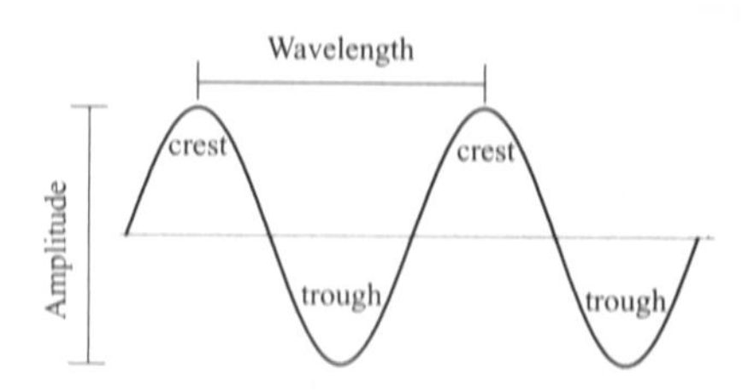

Figure 9.2 The two defining characteristics of a wave are its *wavelength*, which is the shortest distance over which the wave repeats itself, and its *amplitude*, which is the maximum vertical extent of the wave from its highest point to its lowest point.

surfer, or check out a big-wave surfing video online, what's big is the amplitude!

These two features, *amplitude* and *wavelength*, are present in all waves, and their meanings are always the same. Sometimes though, instead of the wavelength, it is simpler to talk of the *frequency*, which measures how many wavelengths we can fit into a fixed distance. Think of the frequency as how "frequently" the crests repeat. For long wavelengths, we can fit less of them in a space, so the frequency is lower, while for short wavelengths, we can fit many more, and so the frequency is higher. There is therefore an *inverse correlation* between the wavelength and the frequency: *the higher the frequency, the shorter the wavelength*. It is easy to appreciate this visually in Figure 9.3 where sine waves with three different frequencies (and wavelengths) are shown. If we know one, then we can deduce the other, so the wavelength and frequency convey the same information and may be used interchangeably.

With just these basic features, we are ready to map people's personalities to waves, as a first step toward understanding how people match up in a relationship. In principle, this is quite simple—with every individual's personality, we can associate a unique waveform. Think of our state of mind, our feelings, our preferences, and our personalities as a superposition of repetitive wave patterns unique to each one of us; those patterns determine who we are. If you are biologically inclined, you could think of all those patterns akin to some sort of brain waves, but the actual biology of it is not relevant or even necessary here.

But how do we represent waveforms for individual human beings in practice? After all, we are pretty complicated creatures, and drawing the waveform for a human being on paper can be a horrendously complicated task. We can imagine that by mapping out all the likes and dislikes, behavioral traits, and eccentricities of an individual, we can draw out some kind of intricate pattern that repeats itself over the course of an individual's life. But we need to have some rules to relate wave patterns to human personalities. How do we find those rules? Fortunately, there is a well-established and systematic way to construct any complicated wave pattern starting with just very simple ones; it is called *Fourier analysis*, after the nineteenth-century French mathematician and physicist, Joseph Fourier, who developed it.

Fourier analysis is essentially the Lego of waves. In a Lego set, we can build any complicated structure we like with just a few different types of building blocks; in quite the same way, we can build any complicated wave pattern by judiciously mixing a bunch of simple waves. Vice versa, we can deconstruct and analyze any complicated wave by breaking it up into its so-called Fourier components—the bits of different contributing simple waves. Here is how it works.

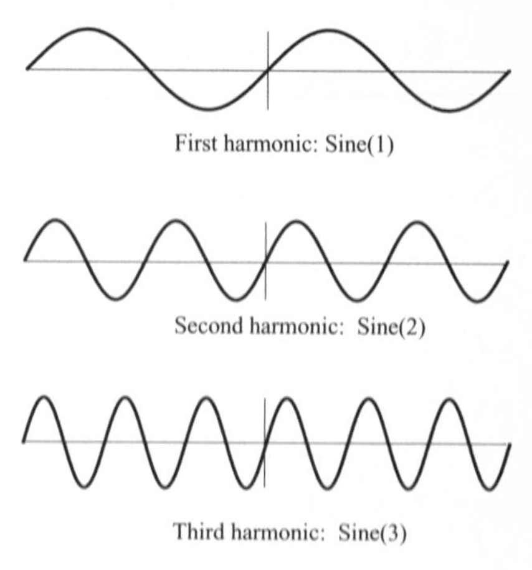

Figure 9.3 Sine waves with different frequencies are shown: Sine(2) has twice the number of crests (hence twice the frequency) that Sine(1) has over the same span, and therefore its wavelength is half as much; likewise sine(3) has thrice the number of peaks (hence thrice the frequency) as sine(1) and therefore a third its wavelength.

We first need to identify the building blocks—the elemental waves—and for this, it makes sense to pick them to be as simple as possible, so it is customary to choose sine waves. To build a whole set of distinct elementals, we use the primary feature of waves: the frequency (or equivalently the wavelength). So our Lego set of waves are just sine waves of different frequencies, sine(1), sine(2), sine(3), and so on, where sine(2) has twice the frequency of sine(1), and sine(3) has three times the frequency of sine(1), continuing thus to higher and higher multiples all the way up to infinity. We can see some examples in Figure 9.3. They are called *harmonics*: the first one—the *fundamental* frequency—sine(1) is the first harmonic, sine(2) the second harmonic, sine(3) the third harmonic, and so on. If all this sounds very musical, that's no coincidence, because these *waves describe musical notes as well*, and we could conjure a more poetic image by dubbing our entire analysis here as the music of human relationships.

With the building blocks at hand, we can now create any kind of wave we like, to match any personality type. To do that, *we associate one specific frequency with every quality of a person*, and we are free to pick and choose them as we like. For example, sine(1) could represent taste in music, sine(2) work habits, sine(3) culinary tastes, and so on. Once we have made the associations, it is just a matter of determining how much of each elemental wave we need to add in to the personality blend, and that just happens to be the amplitude of each component. *If a quality is strong in a person, that harmonic would have a large amplitude; if it is weak, it would have a small amplitude;, and if absent altogether, the amplitude would just be zero.* And for every choice and combination of amplitudes of the relevant harmonics, we will have a different waveform corresponding to a distinct personality. The choices are infinite, as they should be in order to describe infinitely different personalities possible.

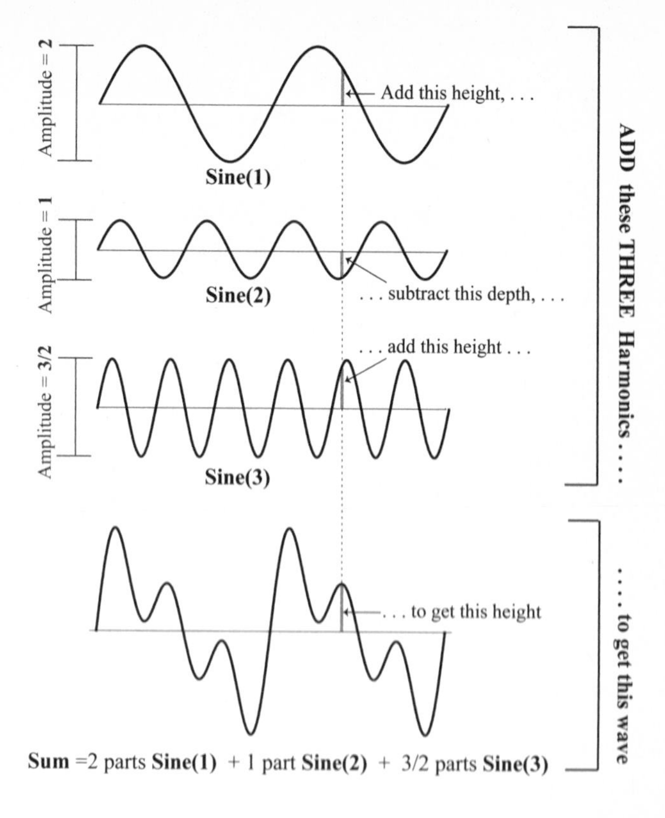

Figure 9.4 Adding waves: Line up the different contributing harmonics (sine waves) one above the other as shown. The vertical height of each harmonic is determined by its amplitude—the bigger the amplitude, the taller the wave. Draw a horizontal median line for each harmonic: the height or depth of each harmonic at any horizontal position (going left to right) is measured up or down from that line. The key principle is that at any position, if all the harmonics have crests, their sum will be an even higher crest, like riding several big waves one on top of the other, but if some have crests and others have troughs, there will be cancellations, and their sum will be reduced. So to find the sum of the harmonics at every position, the corresponding point on each harmonic contributes as follows: If the point is above the median, its height is added, and if it is below the median, its depth is subtracted. An example is shown where the dotted line connects corresponding points: in the sum of the harmonics (shown at the bottom), the height at that particular position (shown as a gray vertical bar) is found by adding the heights of the vertical gray bars in Sine(1) and Sine(3) and subtracting the depth of the vertical gray bar in Sine(2). By doing this for all positions, we get the complicated shape of the "Sum" wave at the bottom. Drawing this by hand can be quite tedious, but, fortunately, computers can do the entire curve in an instant.

This is a lot like when we cook; we have a kitchen full of ingredients, and our recipe might dictate that we add two cups of flour, one cup of egg white, three half-tablespoons of oil, and so forth. For waves, the measure of "how much" is the amplitude of each wave. So, in very much the same way, we can "cook up" a waveform with two parts sine(1), one part of the sine(2), and three halves of the sine(3), and we would say that the amplitude = 2 for sine(1), amplitude = 1 for sine(2), and amplitude = 3/2 for sine(3), and when we add them up as shown in Figure 9.4, the result already looks a bit complicated, and that is with just three components! The figure explains how we add up waves: We line up the waves and, at every position left to right, we add the heights and subtract the depths as measured from the horizontal median lines of all the contributing harmonics. Thus, at positions where all the waves rise up, their sum is higher still, but elsewhere if some waves rise while others fall, there will be mutual cancellations and the sum will be reduced. It would certainly be very tedious to add up the waves drawing them by hand this way, but that's no problem because computers can do it instantaneously!

As we add more and more components for the various facets of human personalities, we will end up with some really complicated waveforms. That is no problem, however, because no matter how complicated it gets, we can always break them down into the simple, individual sine wave components with the Fourier rules. But even without going into all those details, we can already deduce what the wave patterns would look like for a range of different personality types:

An interesting person has a lot of different Fourier components, corresponding to many-faceted interests and character traits.

- A dull, boring person will have relatively few.
- Dominating personalities will have waveforms with high amplitudes, because they have strong qualities that will overwhelm those of anyone they meet.
- Mild, easy-going personalities will have low amplitude wave patterns. They can adjust to personalities with stronger amplitudes.
- Complicated people have complicated wave patterns; they are very tough to match.
- Simple people have simple patterns with few simple components that can be easily matched and analyzed.
- Psychopaths have crazy wave patterns, perhaps constantly changing, very unpredictable, and hard to figure out and hard to match.

You can keep on adding to this list—try it.

However, there is something very important missing that I have not told you about yet. All that I have said so far is well and good for mapping out individual personalities, but *to analyze a relationship between two people, we need to compare their waveforms*. While the frequency identifies a specific quality, and the amplitude tells us how strong that quality is in a person, it is how that quality varies from person to person that matters the most in relationships. For example, consider your taste in music; we pick a particular harmonic to represent it, and strong amplitude would imply that it is an important and defining quality for you, but we do not have a way to distinguish your specific tastes in music. To reflect that information, we need another feature of waves called the *phase*.

The word "phase" can mean a lot of things in everyday life, but it has a very specific meaning in the context of waves. Suppose we have two similar waves, both with the same wavelength but possibly different amplitudes, as shown in Figure 9.5, and we ask, are they "in sync" or "out of sync" with each other? Do they rise and fall together? If they do, then we say they are *in phase*. But if they are out of sync, meaning that the crests of the two waves are displaced relative to each other, then they are said to be *out of phase*. And exactly how much out-of-sync they are is given by their *phase difference*, which is just the distance between corresponding crests of the two waves, as we can see in Figure 9.5. The minimum phase difference of course would be zero, when the waves are perfectly in sync. But there is also a maximum possible phase difference ("out-of-syncness") that occurs when the crests of one wave are *exactly halfway* (or one half-wavelength) in between the crests of the other wave, so that one wave has a crest exactly where the other has a trough like in Figure 9.5 (c).

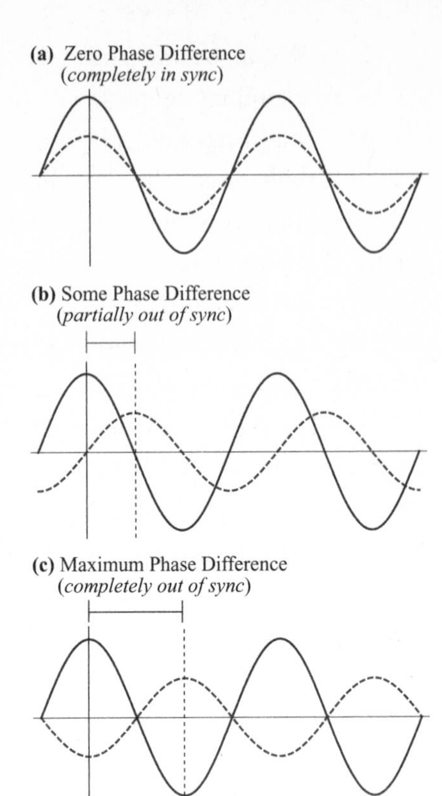

Figure 9.5 Phase of a wave is important for comparing waves. When two waves rise and fall together, their phase difference is zero as in (a); otherwise not, as in (b) and (c). Particularly if one wave reaches its peak exactly where the other one is at its lowest point like in (c), then the two waves have maximum *phase difference*.

The concept of phase has the secret for understanding relationships, because with it we have the means to *compare* the variations of every quality (harmonic) for any two people. A simple example will illustrate. Consider a harmonic that we associate with musical tastes. We will assign a phase of 0 to a preference for classical music, and the maximum phase of ½ wavelength to the other extreme of the musical spectrum, perhaps heavy metal. Everything else—rock, pop, new age, country, folk, jazz, *taiko* drums, *bhajans*, and so forth—will have some phase value in between. For this to work, we will have to carefully arrange similar types of music to have similar phase values that are close by, so for example, new age would have a phase value close to 0, near classical music, while hard rock would be close to ½, near heavy metal. *Phase, therefore, is like a sliding dial* that goes from 0 to a maximum of ½ wavelength as we see in Figure 9.6, and once we have calibrated the dial by assigning phase values for all tastes in music, then your own particular musical taste will correspond to some specific phase value or range of phase values on that dial.

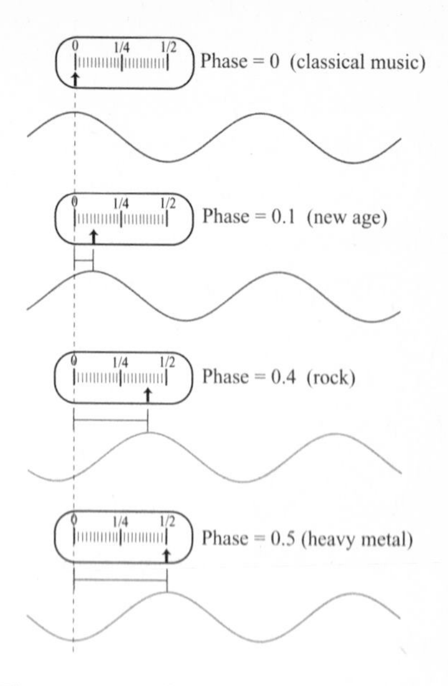

Figure 9.6 Different phases of a wave (measured by how much the first crest is shifted from the reference line) can be calibrated to represent variations of a particular facet of life, for example, musical tastes. In this example, we assume that a taste for classical music is completely out of phase with taste for heavy metal; therefore, the waves representing them have the maximum phase difference.

Now that was just for one component, and we would do this with all the Fourier components (or harmonics) corresponding to all relevant facets of human interests and behavior. The various *phases* associated with them hold the key to relationships, as we will now see.

Say you have been dating someone for some time now, and you want to figure out if the two of you have a relationship future. Wave mechanics can find you that answer: take the wave patterns for the both of you, break them down *à la* Fourier into all the different harmonics, then bring them together, *pair by pair*, corresponding to each facet of your personalities, and match them and add them up. The result will tell you all!

To see exactly what happens, let us track one particular quality, something that is important to both of you, and look at the harmonic associated with it. It could be anything at all, but to be concrete, let us say once again that we are comparing your musical tastes. Now if you both happen to have pretty much the same preferences and tastes in music, then the harmonics for both of you would have the same phase—the crests and troughs would line up as you can see in Figure 9.7(a), so that when you add them together, the combined amplitude gets larger still, the troughs get deeper and crests get higher, and that is a happier, stronger wave. In the language of quantum physics, your musical tastes are in *constructive* interference. As far as music is concerned, you are a *perfect match, because you are in phase*, and your joint amplitude is stronger and more enhanced, and you would say you share common musical tastes. In your relationship future, I see you going to a lot of concerts together, spending evenings listening and even dancing to music together, sharing your musical discoveries with each other, leading to an overall musical enriching of your relationship.

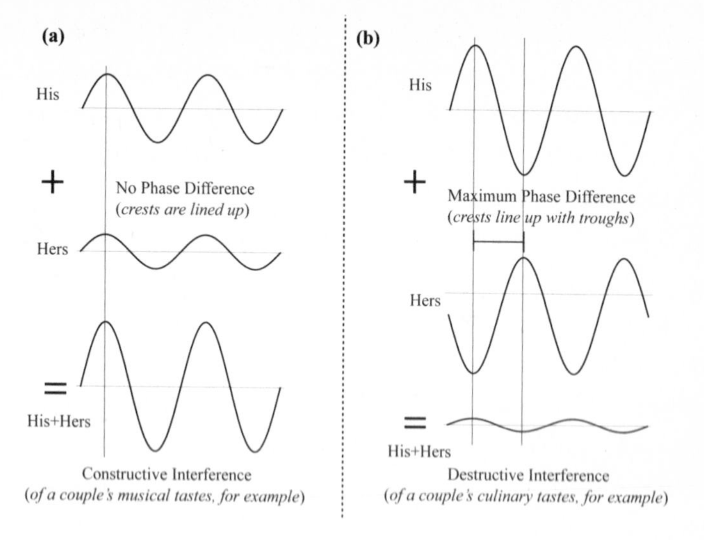

Figure 9.7 (a) Constructive interference of waves happens on adding two waves that are in phase. (b) Destructive interference happens when the waves are out of phase.

Let us now look at another facet, perhaps your culinary tastes. Say you both have a hearty appetite, take delight in your meals, and eat out often, so that the associated harmonics in your wave patterns have strong and similar amplitudes. But unfortunately, your tastes in food are absolutely opposite—you like your meals spicy and exotic, and you think of ghost chili as a treat to look forward to, but you happen to be with someone who is allergic to spices and thinks of boiled vegetables as manna from heaven. This means that your culinary harmonics are completely *out of phase*, meaning that where yours has a *crest*, your partner's has a *trough*. So when you add them together they simply *cancel each other out*, and as you can see in Figure 9.7(b), your combined amplitude is very much diminished. In the jargon of quantum physics, this facet of your relationship is in *destructive* interference. The implication for your relationship is that in being together, you would have to give up on the prospects of ever enjoying a good meal together as a couple, and over time you could go from bemused toleration to utter disgust of the other person's choices of food, and it could be the cause of a serious rift in the long run.

Well, these two somewhat extreme examples should give you the general idea. We would do this with every characteristic or Fourier component of your personalities. The best-case scenario is that all of your components are absolutely in phase or close to it, and your combined waveform will have much stronger amplitude in general, and then you would have found your soul mate. On the other hand, the worst-case scenario is that you and your partner are completely out of phase on everything—you are in total destructive interference. If you are the unlucky woman, he would be the one you had a series of hellish dates with—perhaps you like poetry, Byron is your hero, and he speaks in dirty prose; you are afraid of heights, and he goes bungee jumping and skydiving; you like classical music and go to the opera, and he likes heavy metal and hangs out at the mosh-pit, and so it goes. Well then, you two have a case of "completely out of phase" personality wave patterns.

Those are the extreme scenarios; most couples would be somewhere in between, where their shared harmonics are only partially in phase or out of phase to varying degrees. The degree of being in phase is what is most relevant for any relationship. Of course, for a successful, happy relationship, you do not need to be phase-matched on everything; that is quite unlikely to happen anyway with all the variations in human nature. You can prioritize the qualities most important to you and see how they phase up with those of someone you are with or would like to be with. The more components you are closely phase-matched on, the greater your chances of relationship bliss.

Although I used individual harmonics to illustrate how constructive and destructive interference arise from phase-matching, we could actually bypass the tedious component-by-component phase-matching altogether. There is a much easier quick tell-all about relationships. You can just add both of your personality wave patterns as a whole. If the result is a general increase in amplitude, you are in luck because most of your components are more or less in phase, and you will most likely have an enriching relationship. But, if the overall amplitude is diminished on adding together, because of cancellations at many frequencies, you should plan on going your separate paths, because you are out of phase on most of your components, and you will stifle and restrict each other over time and end up being miserable together.

Wave mechanics clarifies a persistent fallacy about relationships. We have all heard the cliché that "opposites attract." But I always had my suspicions about whether attraction implies eventual success of a relationship, and it turns out that wave mechanics tells us something different. Even if this whole "opposites attract" thing were true, a relationship of opposites is hardly likely to last since from a wave perspective there would be destructive interference all around. However, a more accurate, but related, relationship statement would be that "opposites might attract, but complements last." What I mean by that is that two people who have a lot of complementary personal traits (Fourier components) that are totally absent in the other might find each other attractive, and the resulting relationship could be very enriching and stable, since the relationship would bring in new facets absent in their single lives. Say a woman who is an exceptional artist hooks up with a man who could not draw a straight line to save his life, but unlike the woman, he happens to be exceptionally handy with tools. Then their relationship would be that much more enriching for both, since their joint waveform will gain completely new components that were absent in their individual ones.

Well, there you have it, the elements of a sure-fire algorithm to analyze relationships. With a bit of work, we could easily write a computer program to do all of these. Why limit yourself to twenty-eight or twenty-nine dimensional match-making promises out there? With wave mechanics, you can in principle map *every dimension* of your personality; it is all about how many Fourier components you want to include, and with modern computers, there is almost no limit. All that I have said here is just the tip of the iceberg. In fact, we can understand almost everything about relationships by analogy with waves and not only relationships, but also much about human personalities and compatibilities in general. But that would require a whole other book to develop and describe. However, understanding a problem is not the same thing as finding its solutions, although it is assuredly a great start. Therefore, there is no guarantee that such knowledge will make you immune to future relation-ship-wrecks; after all, we humans are creatures of instincts and habits, with an inclination toward repeating the same mistakes over and over again. But perhaps rational thought and logic, backed by this new perspective, will help a bit the next time around. At the very least, the next time you break up, you can do it in style, and say, "Honey, I love you very much, but I am just not in phase with you," or perhaps depart with the cryptic line, "My love, we are in destructive interference."

CHAPTER 10
RULES OF ATTRACTION

The intangible forces that bring people together or tear them apart may seem quite random and unpredictable—that is, until we look at their analogs in nature that hold the universe together, from atoms to galaxies. The rules that govern those forces of attraction (and sometimes repulsion) reveal patterns that underlie the forces at work among people as well.

If all relationships started with an objective compatibility analysis, there would be very few bad relationships. Matchmaking and dating sites claim to be doing just that, but you can be sure that not all their "success stories" lead to living happily ever after. One obvious flaw is that they have to rely upon client-provided data, which are seldom objective or entirely truthful. Never mind those dating services—with reliable personality profiles on hand, we could use the wave mechanics from Chapter 9 to determine compatibility and project long-term success of a relationship. But, no matter what method we use, there is a serious catch: the timing! Do we ever get the opportunity to do any such analysis when it matters the most? That would happen to be before we get into a relationship, when we still have a fighting chance to exit the bad ones. Usually not.

Why? Because relationships are almost never planned with genuine compatibility in mind—yes, not even in arranged marriages still popular in parts of the world (planned and arranged they might be, but usually with very different priorities, like "will the in-laws get along?" or "what's the size of the dowry?"). Generally, any analysis, if it ever happens, is forced upon us *a posteriori* when things have already gone to the dogs. So, unless some match-making service was involved in getting you started, you won't get to seriously analyze your compatibility with someone until you actually get to know the person—perhaps a bit too well. But by that time, you have already long succumbed to the initial attractions; you are already way too deep in the relationship. We all take a chance every time we act upon an attraction for someone because we can never know for sure whether we would click and get along well. The consequences are all around us to see: A lot of people stay together in terrible relationships, even when they have almost nothing in common or, in the language of wave mechanics, even when they are in complete destructive interference at most frequencies.

Attractions and attachments are seldom rational and never within our complete control. With so much irrationality built into human attraction, it is usually not compatibility that brings people together in the first place, and it certainly cannot explain why people stay together even when they are wrong for each other. There are other forces at play—forces of attraction that are often too strong to allow any logical decisions on our part. Perhaps we would have a chance to conquer and overcome those forces—if we could figure out how they work. Thus, our goal in this chapter is to understand the rules of attraction of the forces that bring people together. We can largely bypass the irrationalities of the human mind when it comes to love and attachment, because many of the rules of attraction (or repulsion) we operate by are actually not that different from those that hold the physical universe together.

Let us start with an extreme scenario—that of fatal attractions—relationships that you just can't get out of, no matter how hard you try. Sometimes it can get quite nasty and psychotic; you could be manipulated, threatened, stalked, and abused. But more often than not, it is you yourself who are bound by an obsessive attachment, love, or lust, or just a deep-rooted emotional dependency you can't break free of. Well, nature has its analog for such a situation lurking in the depths of space, something really extreme, forces of nature gone wild: *black holes*. It is hard to miss them in popular media, movies, or science shows on television, and much of what they say is actually true. Nothing can escape a black hole, not even light, hence its name—black holes manifest as invisible regions of complete darkness, like holes punched in space. But what is less well-known is that every object has the potential to become a black hole, just as every relationship has the potential to spiral into an emotional and mental trap.

If you squeeze down any object to extremely small dimensions, it will eventually become a black hole. But I mean squeeze down really, really, really tight; for instance, the earth has to be squeezed down to the size of a peanut to transform it into a black hole. Every black hole has something called an event horizon—a boundary in space-time around it, which acts like a surface of no return. Outside and far away from this event horizon, it's pretty much business as usual. So even if the earth is squeezed down to the size of a peanut, the moon orbiting around the earth will continue to feel the same gravitational force. The trouble begins if you were to creep close to the peanut-sized earth's event horizon—if you cross it, you will be sucked in, and once you have crossed the event horizon, you will never get out again.

That's very much like those bad relationships. Far away from the sphere of influence of the fatal attraction, you are safe, but as you get emotionally and physically closer to the person, it becomes harder to escape, and once you cross the relationship equivalent of an event horizon, you are doomed—you may never get out again into the peace and tranquility of your known universe where you used to be happy and carefree. Nature's message here: If we suspect that we are getting close to a "black-hole" relationship, the only solution is to get the hell out as fast as possible because if we get in too deep, there is no way out, perhaps ever.

In our universe, black holes come in all forms and are classified according to size. The really nasty ones are the super-massive black holes, lurking at the center of many galaxies, including our own Milky Way galaxy. Such black holes can have the mass equal to the combined masses of billions of stars like our sun. And there are people like that, with egos the size of a million suns backed by fatal charm that lure in weaker individuals. Cult leaders are examples that come to mind. They do not just affect one individual; they can trap and disrupt the lives of hundreds of people unfortunate enough to come into their sphere of influence; they are super-massive ____holes at work—you may fill in the suitable descriptive noun of your choice.

Fortunately, most relationships are not that extreme, and neither are the forces of nature most of the time, but the similarities persist even under normal circumstances. For one, many a romantic relationship of great promise and potential has fallen victim to distance and separation: "Out of sight, out of mind," as they say. Long-distance relationships are hard to maintain, and most inevitably fall apart if the separation is sustained for long. Before you go off and underplay the risks of separation with some romantic excuses such as "love will see us through" or "if it could not survive separation, it was not strong enough," you might want to consider that, in this matter, relationships are susceptible to a law of nature, equally true in the physical world: Most of the fundamental forces of attraction that hold the universe together fall off rapidly with distance, as well.

Among the forces of nature, gravity is the one with which we are the most familiar. After all, it is our constant companion, and if you are ever in doubt, try climbing the stairs to the top of some tall building in your neighborhood; gravity will rudely intrude upon your thoughts long before you have made it to the top. But gravity is not just keeping us all grounded, attached to the surface of the earth; it is also the large-scale glue of the universe. Gravity holds the planets to stars, stars to galaxies, keeping everything from drifting away at random from one another. (Well, the universe is still expanding as a result of the Big Bang, the primal explosion that started it all, so despite gravity, galaxies are still drifting away from one another, but without gravity, there would be no galaxies or stars to start with.) Black holes do their attracting with gravity, as well—it is just the extreme high-end limit of gravity.

Gravity obeys what is called an inverse square law, discovered by Sir Isaac Newton in the seventeenth century. He was also the first to recognize the existence of gravity itself as an invisible, ubiquitous force. The idea, so the story goes, fell upon his head in the form of an apple from a tree, which apparently triggered the thought that the force at work on the apple is the same force that keeps the moon forever circling around the earth. So how does this inverse square law work? Every material thing in this universe attracts everything else with this force of gravity. The more massive the object, the bigger its pull, so the earth can keep the moon moving around it, because it has a big enough pull. You and I, on the other hand, cannot rely on our sheer gravitational power to attract a romantic interest; our force of gravity just is not strong enough, because our mass (or weight) is small. But perhaps there are exceptions; after all, big strong hunks tend to attract more women, so perhaps it is just that extra gravitational edge that comes with bigger bulk. Who knows?

But why is this called an *inverse square law*? This is because if we increase the distance between two objects to twice their original separation, the force between them goes down by a factor of four, which is two times two, or two squared, or move them apart three times farther, then it goes down by a factor of nine, which is three times three, or three squared, and so on. The "inverse" comes in because increasing the distance has the opposite or inverse effect of decreasing the attraction, and "square" because that reduction goes as the square of the distance.

The surprising thing about the inverse square law is that it is not just true for gravity; it is also true for the other forces in nature we all have some experience with: the forces of electricity and magnetism. Besides these, there happen to be two other forces in nature, the strong and weak nuclear forces, which work at subatomic levels and are manifest in atomic bombs, nuclear reactors, and particle accelerators, but we never encounter them in everyday life.

How is it that forces that seem so different—gravity, electricity, and magnetism—are dictated by the same rule? There is actually a very simple visual way to understand why that is so. Let us visualize the force exerted by an object as a bunch of uniformly distributed lines reaching out from the object, sort of like the spines on a sea urchin or the quills of a hedgehog as shown in Figure 10.1. Now, imagine a spherical surface surrounding the object—two such surfaces of difference sizes are shown in Figure 10.2. All the lines of force penetrate such a surface. The total number of lines remains the same no matter how large we

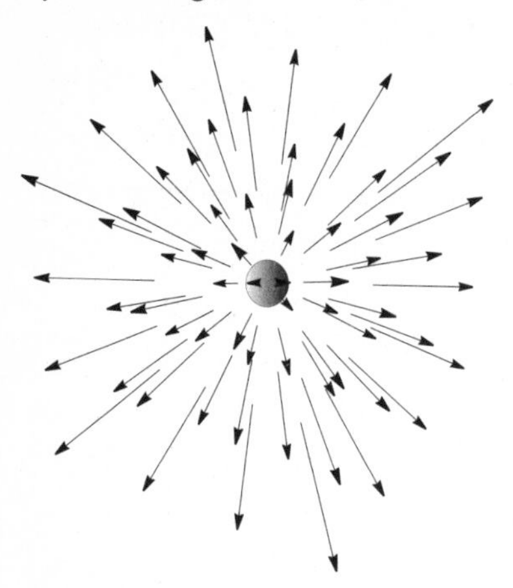

Figure 10.1 Lines of force are visual depiction of force fields due to gravity or electromagnetism. The force becomes stronger closer to the source, since the lines get closer together, and their spatial density increases.

choose to make the enclosing sphere, but as the sphere gets larger, the lines would get increasingly farther and farther apart. As a result, the same square bit of area on a larger and farther out surface will have fewer lines penetrating it than if it were closer, as can be seen by comparing the two spheres in Figure 10.2. But the surface area of the sphere increases as the square of its distance, r, to the object at its center.[13] This means that the density of the lines of force will decrease as square of the distance as we go farther out from the object. We can visualize the force as being proportional to the number *density* of these "lines of force" at a given distance from the source; therefore, the force must *decrease as the square of the distance as well.*

The inverse square law is therefore simply a consequence of the fact that we live in a three-dimensional world, hence the sphere. (In a two-dimensional world, we would have drawn a circle, and we would have arrived at a different law based on the formula for the circumference of circles.) That is why all of the macroscopic forces of attraction—gravity, electricity, and magnetism—follow the inverse square law, because we experience them all in three-dimensional space.

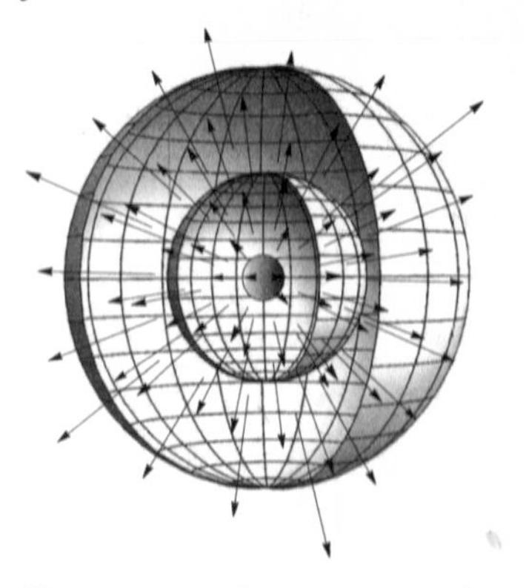

Figure 10.2 Geometric explanation of inverse square law: The force is proportional to the density of the lines of force. On moving out farther from the source, the density decreases, as can be seen: the number of lines remains the same, but the surface area they are spread over increases as the square of the distance from the center; hence, the force decreases inversely to that.

As with most natural forces of attraction, separation weakens relationships. But strength

[13] The surface area of a sphere with radius (distance of the surface to the center) of length denoted by "r" is given by the formula $4\pi r^2$, where the Greek letter π is a constant with the approximate value of 3.1415.

of relationships might not fall off precisely as the square of your separation, because the geometry of human dynamical space is not necessarily three dimensional. In fact, we might think of the forces of human attraction as multidimensional—every defining aspect of it mapping out a distinct dimension. And the higher the dimensionality, typically the more rapidly the forces of attraction fall off with distance.

But the real trouble with separation is not just the weakening of the attraction, but the likely possibility that the separated ones will come within the sphere of other more immediate attractions, which will inevitably have a stronger appeal due to their proximity. What happens is very much like the gravitational tug of war among competing stars for their planets. If in the course of travel through interstellar space, our solar system were to pass by a star much more massive than the sun, the bigger star could very well pull all of the planets, including Earth, into orbit around itself—farewell, poor old sun. That could be poor old you, if you happen to lose your darling to the stronger attractions of a charming renegade during an extended trip, say in some distant exotic country, far away from you. Extended long-distance separations can spell serious relationship trouble—you can't mess with the laws of nature!

There are exceptions. Sometimes an occasional bit of separation seems to strengthen the relationship, by helping people get some honest perspective and overlook the minor perturbations that get magnified with everyday familiarity. In fact, there are some folks who can only seem to connect with their loved ones with a bit of spatial separation; you appreciate your spouse a whole lot more when traveling, all by yourself in a strange bed in a lonely motel room, but when you are in town sparks fly and even the pets run for cover. Nature has an analog for such chronically stronger-with-distance kind of relationships, as well: the strong nuclear interaction.

This is one strange interaction, locked forever within the heart of atoms, something we can never see at the scale of everyday life. This is the force that binds quarks to quarks to form protons and neutrons—the building blocks of atoms. Unlike the forces like gravity and electromagnetism we experience in everyday life, the strong nuclear interaction does not decrease with distance. This has some very strange implications: If we try to pull quarks away from each other, their binding force does not decrease, so we can never get a free quark—they are always bound together either in pairs, called mesons, or in triplets, called hadrons (protons and neutrons are hadrons). What is even more crazy is that if we pull two quarks away from each other, the energy of their mutual attraction becomes stronger and stronger, and at some point, there is sufficient energy to create two more quarks in between, *à la* the mass-energy equivalence of Einstein's $E=mc^2$. Not only can we never separate two quarks, but if we try to pry them apart, we just keep on creating more in between. But the strangest feature is that their attraction is relatively weak when they are really close together. So in close proximity, a pair of quarks behaves as if the other one does not exist as far as the strong nuclear attraction is concerned. This is called *asymptotic freedom*. For sure, some relationships can certainly seem to be asymptotically free, too, particularly after being married for a long time!

Well, most relationships are not; they are more likely to be of the fall-off-with-distance kind, like gravity. And for something like that, Einstein's picture of gravity is a more dramatic and visual one. In his *theory of general relativity*, a massive object essentially distorts space and time around it. It is quite impossible to visualize such a distortion in reality since we already live and exist in three-dimensional space. Instead, we can get a sense of it by looking at a two-dimensional (2D) analog. Think of a very tightly stretched rubber sheet with square grid-lines marked on it as a 2D space. Now, if we put a marble in the center, the sheet will pucker down and distort at the position of the marble, as shown in Figure 10.3. If instead of a marble there is a more massive object, the distortion will be bigger. Now, if we roll a coin around that pucker, it will revolve around the marble, just like the moon does around the earth. In the extreme case of a super heavy mass, the sheet will just be punctured and form a hole—like in those gumball vending machines in the mall or the supermarket, where an inserted coin spirals round and round and eventually falls down into the hole in the center, like in Figure 10.4—that's pretty much what would happen to an object shot sideways into the event horizon of a black hole. Quite literally, black holes punch a hole in space-time, referred to as a *singularity*.

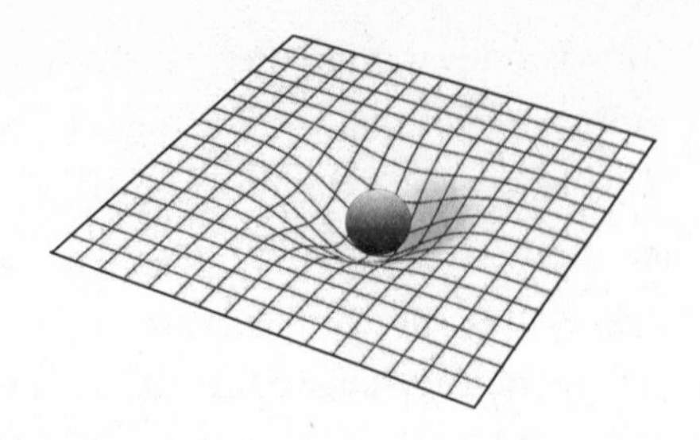

Figure 10.3 A two-dimensional visualization of how gravity distorts space-time around it. Gravitational attraction can be understood as arising due to an object moving toward the "depressions" in the warped space-time.

Actually, this view of gravity provides a much more telling analogy of why we are attracted to certain people. They distort and morph the social space around them, and we fall into that distortion and end up in their social orbit. The stronger the attraction exuded by such a person,

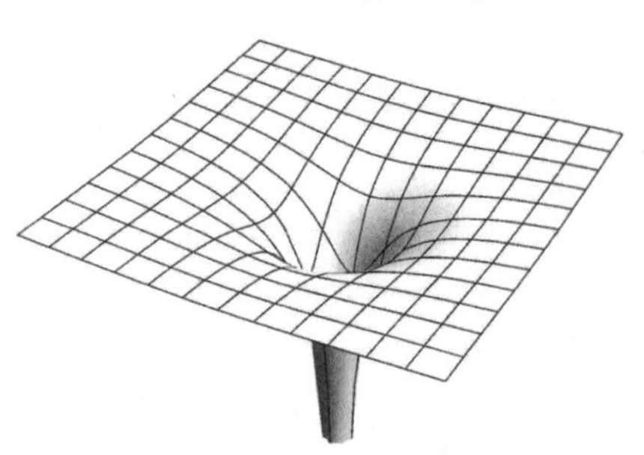

Figure 10.4 If the gravitational distortion is very strong, it can puncture a hole in space-time called a singularity; this is what happens in black holes.

the deeper the well of social vortex around them, an
is to pull free, and in the extreme case they can bec
equivalent of black holes puncturing a hole in the
in their vicinity. Think of the popular kids in your high school; they had (or have) a lot of people in their social circle, didn't they? Wherever they go, any party they are at, there is always a group of people around them in their warped social space.

In life, many of us still have some romantic faith in the thought that "absence makes the heart grow fonder." Certainly, for some couples who have been together for a while and are starting to take each other for granted, an interlude of separation might be a relationship tonic. Perhaps all relationships have a degree of asymptotic freedom at very close proximity. Nevertheless, long periods and large distances of separation will eventually make the heart grow colder rather than fonder as people get pulled into the orbit of other proximate and immediate attractions.

CHAPTER 11
NEWTON'S LAWS OF HUMAN DYNAMICS

> *More than three centuries ago, Sir Isaac Newton wrote down three dynamical laws that define the motion of everything we know of—from baseballs to the stars. These same laws can explain much of human dynamics, as well, from our inherent resistance to change to our often predictable response to the forces at work in our lives, and even our expectations of justice and retribution.*

We take comfort in what is familiar, we are usually slow to change, and with the increasing weight of years and experience we become even more resistant to changes in our set trajectory of life. Yet our lives are defined by changes, however reluctantly we embrace them, and those changes are often brought on by external forces not completely in our control. Biological forces transform us physically, career pressures move us and take us to places we never expected to be, and attachments and family ties push and pull us toward difficult decisions. And through it all, our actions in life always have some consequences that often come back and affect us in various ways.

I need not tell you that the lines above refer to people and how we live our lives. But, believe it or not, with a few simple changes, all the thoughts in the above paragraph could easily belong in a textbook of physics—take a look:

It is the nature of all matter in the universe to resist changes to its state of motion or rest, and with increasing mass or weight, they become ever more resistant to changes in their set trajectory. Yet change defines the physical universe, and those changes are usually brought on by unbalanced forces. Nuclear force can transform the elements, gravitational forces move things physically often taking them to distant places far from their place of origin, and electromagnetic forces push and pull at objects with the choice of attraction and repulsion. And underlying it all, every action always has an equal and opposite reaction.

To understand just how this could be, we need to go back to the seventeenth century, when Sir Isaac Newton almost single-handedly laid down the rules that explained the dynamic mechanism of the physical universe. Those rules defined our scientific worldview, absolute and unchallenged, for the next two and a half centuries; they still continue to be directly relevant today, and they always will be. It's true that some very smart predecessors like Galileo, Copernicus, and Kepler laid the groundwork—the "giants on whose shoulders he stood to see farther," as Newton famously acknowledged, with uncharacteristic humility. But really it was Newton who brought it all together and created a single, coherent, and complete worldview that explained how everything moved not just here on Earth, but in the heavens as well.

Newton wrote down three simple laws that pretty much define how everything moves—everything that we are likely to encounter in our lives—from motor cars, planes, and rockets to the motion of the earth, the moon, and the planets. We can accept such ideas quite casually these days, without as much as a raised eyebrow. But in

Newton's time, it was an intellectual upheaval of titanic proportions for someone to dare say that the rules that apply to the terrestrial sphere of sinful and impure humans could even be remotely similar to the rules that govern the supposedly pure and pristine heavens. Well, Newton not only said it, but also proved it to be so, and today every successful mission to the moon and the planets validates the universal applicability of his laws, over and over again. Our perspective about our place in the universe has never been the same since.

Newton's ideas pretty much laid the foundations for the entire field of physics. It is true that in the early twentieth century, relativity and quantum mechanics modified the Newtonian picture significantly. But for most practical purposes, Newtonian physics is sufficient, because quantum mechanics is truly manifest only at the subatomic level, and corrections due to Einstein's theory of relativity become relevant only when things move extremely fast, at speeds comparable to the speed of light or in the vicinity of extremely massive objects much larger than the sun—and nothing we ever normally encounter is either that massive or travels at such speeds. We inhabit a Newtonian world.

In the centuries since, Newton's ideas also had profound influence on philosophy and humanism and our broader perspectives about our role and place in the universe. Notions of determinism and the absolute confidence we have today that the universe is understandable and even predictable with science, mathematics, and rational thought can be directly traced back to Newton. But even with all that, I doubt that either Newton or any of his contemporaries ever thought that his physical laws could be used to help us make sense of the nuanced nonphysical aspects of the human experience, as well. But, all three of Newton's laws have very direct and important implications in our lives. Let us now see how.

Newton's first law of motion is all about inertia, a central concept in Newtonian physics. Inertia is the natural resistance everything has toward changes. Every material object, big or small, in the universe has a built-in inertia that resists any attempts to change its state of rest or of motion that it happens to be in—meaning that if the object is at rest, its inertia defines how hard it would be to get it to move, or if it is already moving, its inertia defines how much effort would it need to change its direction of motion, to slow it down, or to speed it up. Well, I do not need to tell you that we humans have a lot of inertia built into us, as well. We are all living it every day. Newton's law of inertia certainly applies to us, but in more ways than in the pure physical sense.

The concept of inertia has always been a part of human existence long before it gained currency as a physical concept with Isaac Newton. In life, we prefer to be at rest unless impelled by some motivating force, usually an external one: we are forced to get to work by a certain time in the morning, which drags us out of our state of rest that we would have liked to perpetuate within the confines of our beds; we are forced by thirst and hunger to abandon the state of rest on the couch to get active in the kitchen; we are forced to work at paying our bills by the threat of creditors and collection agencies. On the other hand, if we happen to be on a roll, we tend to continue until compelled to stop—for instance, if you are settled into your steady and stable path in career, neighborhood, and family, you will need strong forces to make you deviate from that path. And that pretty much is the content of *Newton's first law of motion*: "A body at rest tends to be at rest, and a body in uniform motion tends to continue in uniform motion, unless acted upon by a force." He might as well have been talking about human nature.

In Chapter 4, we talked about the universe being intrinsically lazy just like people; well, here is just another natural law validating that. People, just like any inanimate object in the universe, always seem to need some motivating force to change course. Forces of circumstance need to arise for people to change course in life. In the absence of such forces, we simply continue unperturbed and uniformly along the trajectory of life that we happen to be on. Really, life is often simply a battle to overcome inertia.

Newton's second law of motion is the centerpiece of the Newtonian view of the universe, which is captured by the simple relation

$$Mass \times Acceleration = Force$$

It can also be written in another perhaps more revealing way:

$$Acceleration = \frac{Force}{mass}$$

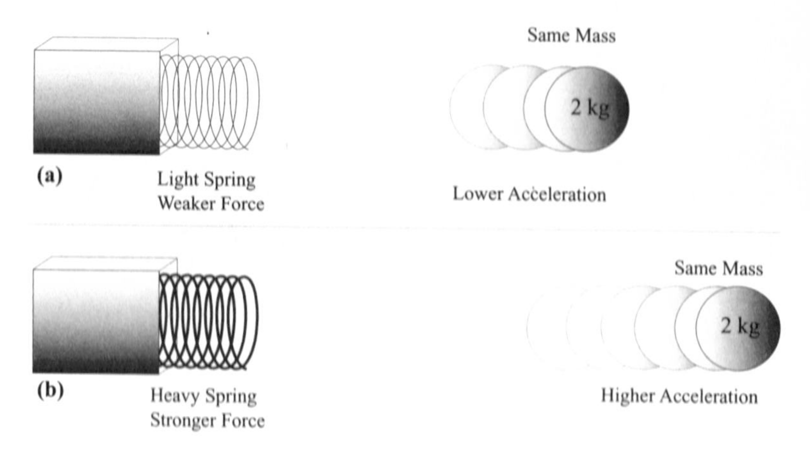

Figure 11.1 *Newton's second law of motion* states that acceleration is proportional to the applied force if the mass or weight of the object is unchanged. (a) A light spring exerts a weaker force and leads to smaller acceleration. (b) A heavy spring exerts a stronger force and leads to a higher acceleration so that the object moves farther in the same time.

Although it seems mathematical, we can ignore the math, because the meaning of this can be understood in common sense intuitive terms. Let us first be clear about what these three words mean, and nothing could be easier because they mean exactly the same thing in physics as they do in their common usage in everyday life. *Force* is what we apply on things or on people to make them move or behave the way we want them to. *Mass* is the amount of matter contained in an object and is proportional to its weight—thus a heavier object has more mass than a lighter one. *Acceleration* is the change in velocity or speed—like when we increase the speed of our car—that is why the gas pedal is called the accelerator.

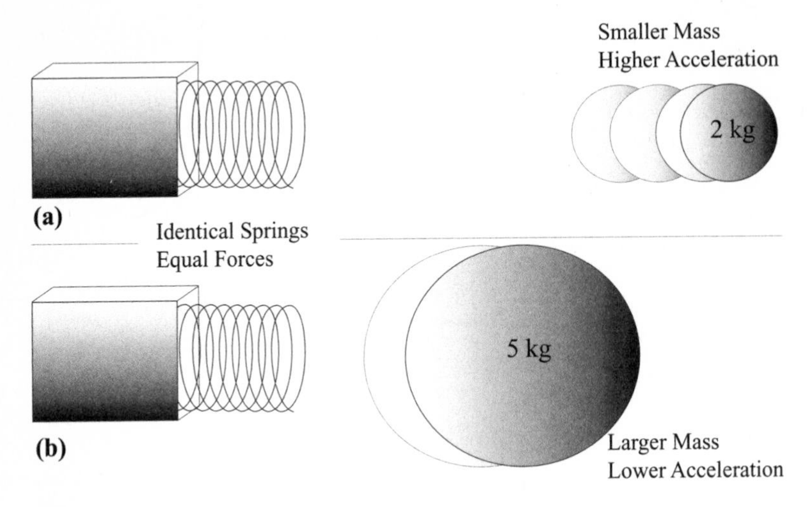

Figure 11.2 *Newton's second law of motion* states that if the applied force is the same, the acceleration decreases as the mass or weight of the object increases. Here, identical springs apply the same force to two objects. For the same time of travel, (a) the lighter object accelerates faster and travels farther, and (b) the heavier object accelerates slower and covers less distance.

Therefore, *Newton's second law simply states that the acceleration that an object undergoes is directly proportional to the force applied to it and inversely proportional to its mass or how heavy it is.* Interpretation: An object will accelerate (or speed up) more (1) if we apply a larger force as shown in Figure 11.1 or (2) if the object is lighter (has less mass) as shown in Figure 11.2. But we already know this from experience. That acceleration requires force should be pretty clear if you have ever driven a car. To make it go faster and faster, you need to apply a force, and that is exactly what you do when you press down on the gas pedal in a car; the engine is churning faster, applying a larger force, making the car speed up. The gas pedal controls the force by controlling the rate at which the gasoline burns. We often see the word "torque" when reading about performance of cars—well, torque is just a measure of the force that makes things rotate or spin, like the wheels of a moving car. A Porsche 911 can accelerate from 0 to 60 mph in close to four seconds because its powerful engines provide a much higher torque compared with the typical family minivan! Besides, the larger mass of the minivan also drags it down. Now, if you have any doubts about how mass affects acceleration, just ask yourself if you were to push a Mini Cooper and a tractor trailer in turn, which one do you think is more likely to start moving, or accelerate, from its state of rest?

Newton's second law has a broader interpretation if we recall that for any physical object, its general position at any given time and/or how fast and which way it is moving are the basic characteristics that define its state in the quantum mechanical sense that we described in Chapter 2. Therefore, acceleration in Newtonian physics is really a measure of the change of state of an object. Well, people have many things that define their state, and so the analog of acceleration in human terms would be the change of state—mental, physical, financial, professional, emotional, familial, and everything else—that defines a person at any stage of life. Forces are literally acting on us all the time to change our state; we all know that. Physical motion requires physical force, but socioeconomic, emotional, professional, and other such changes relevant to us in life are wrought by corresponding nonphysical forces. That generalized definition of a force pervades our language already, and we talk about socioeconomic forces, emotional impulse, and familial dynamics, all of which conjure up imagery of people and society being accelerated along different life trajectories by external forces just like a physical force can accelerate a material object.

Newton's second law states that the effect of a force on an object is determined by its mass[14]—the larger the mass, the greater the force necessary. As we generalize to human and social notions of force, we also have to broaden the concept of mass. We need to understand mass in its broader sense as a measure of inertia of an object. For individual humans, that *measure of inertia is age and experience.* It is well known that as we "amass" experience and the wisdom of the ages, we become less susceptible and less receptive to changes. Young people can, and often do, change their situation rapidly and are much more susceptible to external forces, but older people do not and cannot change course as easily. It would indeed require significant forces to alter the trajectory of a person in midlife—to change horses in midstream—but even small interpersonal forces can easily alter the life trajectory of people in their teens and twenties. For example, while peer pressure is always a significant force all through our lives, it is the most potent in our vulnerable formative years (when we have less mass of experience and constraints), and we yield to it more easily and more often. As we *amass* more experience and burden of various constraints in life, we get experience-heavy, and we become less susceptible to the forces of peer pressure and are less likely to start acting or moving on courses of action just because of what others

[14] We often use the words "mass" and "weight" in a similar sense, and it is fine as long as we are close to the surface of the earth. Our weight is actually a measure of the gravitational pull of the earth on us, and if we go far enough out into space away from the earth, we can be "weightless" as that force diminishes. But we will never be "massless" because mass is just a measure of how much matter we are made of. Newton's second law implies that the force of gravity or the weight of any object is proportional to its mass.

around us say and do. Just as the weight or mass of a physical object is the measure of its inertia; in human terms inertial mass is measured by age and experience. It is borne out by the fact that we generally associate fickleness with youth and stability with age. But there is a strong positive side to the easy changeability of youth as well: Young minds are more flexible and open to new ideas and thoughts, and quite often in history great ideas and great changes came from relatively young people. In the light of all this, Newton's second law provides the most important law of human dynamics:

Rate of change of state of a person is directly proportional to the forces (in the broader sense) acting on that person and is inversely proportional to the amassed years and experience of the person.

This is truly a fundamental law of human dynamics almost as much as the original Newton's second law is a fundamental dynamical law of the physical universe.

Newton's third law of motion has perhaps the deepest implications for our lives. Ever had the reason to mutter, "No good deed goes unpunished"? Or cry out in outrage and frustration, "What goes around comes around"? These are merely consequences of the bigger pattern of life that most of us believe in—that everything we do has some sort of a consequence or reaction on our lives. There is always payback! Our faith in justice is based upon that. Even our religious faith is founded upon the ultimate payback on Judgment Day or in the hereafter in Heaven or Hell. All of this is pretty much the content of *Newton's third law:* "Every action has an equal and opposite reaction." Our lives are more complex than the systems Newton considered, so the reactions are not always exactly equal or opposite and might not be immediate; nevertheless, there is a reaction for every one of our actions, which usually catches up with us eventually. The whole notion of "karma" is really just Newton's third law applied to life.

The law of action–reaction, although simple, is often misunderstood. In physics, the action and the reaction refer to forces acting and reacting. Therefore, the third law implies that for every force, there is an equal and opposite force. Now, here is something that puzzles many students when they learn this for the first time: According to Newton's second law, there needs to be net unbalanced force acting on a body to make it accelerate. But if every force comes with an equal and opposite force, how can there ever be any acceleration? Wouldn't all the "equal and opposite" forces simply cancel out—like two equally strong guys pushing or pulling an object in opposite directions? Wouldn't the net force be zero then, and nothing would ever accelerate, and anything at rest is forever doomed to be at rest? But we know that is not the case, because things do move. Yet Newton's third law is a completely valid and respectable law. So what is the explanation?

Well, the fact of the matter is this: The action and the reaction *never act on the same body*. So, when two pool balls collide, as shown in Figure 11.3, the action of *ball 1* acts on *ball 2,* and vice versa, the reaction of *ball 2* acts on *ball 1*, and because the action and the reaction forces are equal and act oppositely, the balls fly off in generally opposite directions.[15] The action–reaction paradigm applies in every scenario where things move and accelerate. Suppose we are on a treadmill. Our feet push back on the treadmill, and that is the action that causes it to move backward; the treadmill pushes back on our feet with a reaction force and causes us to move forward. Obviously, we can accelerate on a treadmill—we are usually running fast shortly after starting from rest. But the treadmill belt accelerates, too, moving equally fast in the opposite direction as a reaction. With a treadmill, it is easy to visualize the effects of the action

[15] They would move in *exactly* opposite directions if the collision were head-on. But if as shown in Figure 11.3, if the balls collide at an angle, they will bounce off at an angle still in generally opposite directions.

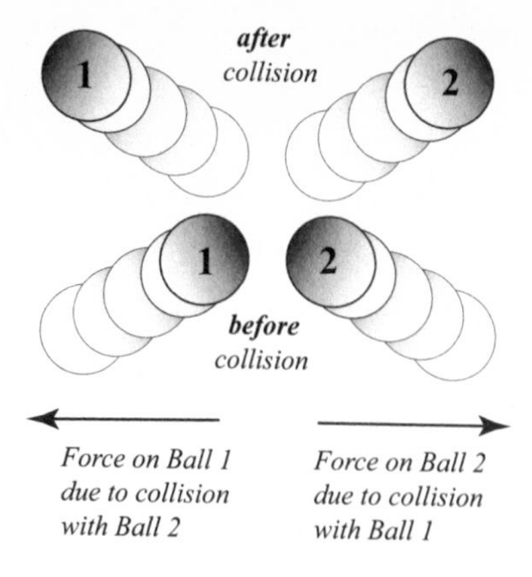

Figure 11.3 *Newton's third law of motion* states that, "Every action has an equal and opposite reaction." If two objects collide with each other, the force of each object on the other one makes them both move away in opposite directions.

and the reaction, but the same thing happens when we walk on the ground; our feet are pushing back on the earth, but the earth being so massive cannot be moved by the puny force of our legs pushing back (recall Newton's second law). This action and reaction thing is what makes it so tricky to operate things in space: If the astronauts push at anything, it pushes them back and they fly off because they are not held by Earth's gravity or by the forces of friction that prevent us from floating and sliding around.

That is pretty much how it works in our lives, as well. Our actions impact others around us, and their reaction is what comes around and acts on us sooner or later. Our actions affect others, and their reaction eventually affects us. What goes around does often come around; we can usually count upon that—Newton's third law at work. You could even make money off the notion of action and reaction, as one shrewd investor named Roger Babson did in the early twentieth century. He applied this whole idea of action and reaction to the stock market and made a lot of money from it, and even got a college named after him. He assumed that what goes up comes down and vice versa, and the market usually does, even in the best and the worst of times. So it is the simplest rule of investing, yet the hardest to follow, to buy when prices are going down, because the reaction will eventually bring them up.

CHAPTER 12
QUANTUM THEORY OF SOCIAL INTERACTIONS

The quantum view of interactions at the level of subatomic particles is remarkably similar to how interactions happen among people—via exchange. The quantum connection reaffirms the key ingredient necessary for meaningful interactions.

What would life be without interactions? It just would not be worth living. We may all crave solitude once in a while, and talk wistfully about getting *far, far away from the madding crowd*—but never forever, because we just can't be by ourselves for too long. Sooner or later, we will feel the need to interact with other people. That is true even for the greatest misanthropes and hardened souls out there. Thus, in these more civilized times when physical torture is considered barbaric, if society wants to punish people who are already incarcerated in prisons, it resorts to solitary confinement. Even a few weeks can subdue tough criminals and gang-bangers of negligible sensitivities, and years with no human contact whatsoever would reduce most people to madness. In fact, interactions are essential for us right from the get-go—it is well known that in the early developmental stages of infants, interaction with people around them is absolutely essential for the proper growth and development of the brain.

Interaction is the basis for all human social behavior. Without interactions, there would be no society, no group dynamics, no fun and festivities, and certainly no culture and civilization; we would each scurry from birth to death in a sad, solitary existence. But if anything, interactions are even more important in the physical world, for without interactions, the universe as we know it simply would not exist. There would be no galaxies, no stars, no planets, and therefore no life; there would not even be stable atoms without the various types of interactions to bind all the subatomic particles together. Something that fundamental in both the physical and the social universes can hardly be without deep parallels. So there are, and in this case, it turns out that while much can be learned about human interactions from their analogs in the natural world, even more can be learned about interactions in nature from the way people interact.

So, what do interactions mean in physics? We say two objects are interacting if they exert a mutual influence on each other, often accompanied by some sort of exchange between them. That's not unlike what happens when two people interact. Thus, the sun interacts with the earth, because the force of gravitational attraction acts between them. Two balls on a pool table interact with each other when they collide. In nature, as with people, there are many different forms of interactions, and in either case, interactions always make everything much more interesting and much more complicated and dramatic. In fact, there would be no drama without interaction. It would indeed be a pretty dull theater play or a movie that has just one actor in it or one in which people just ignore each other. Interactions are the spice of life, just as interactions make the universe tick.

Our ideas about how things interact in the universe have evolved quite a bit over time, and what is really fascinating is that our most up-to-date and current quantum mechanical understanding of physical interactions is remarkably similar to the nature of human interactions—much more so than the classical view that scientists held on to for almost three hundred years, well into the early twentieth century. Actually, those old ideas still work quite well for predicting the effects of interactions, even good enough to place a person on the moon. But when it comes to explaining how and why the intangible and invisible interactions, such as gravity and electromagnetism, work, those old ideas completely fail. It's similar to how we can learn to drive a car without knowing anything at all about how the car works. Our views about natural interactions pretty much stayed that way until quantum mechanics came along—we just used the formulas and gave up on asking *why* for a bit.

So what was the classical view? Since Sir Isaac Newton came up with the theory of universal gravitation, the explanation, or the lack of one, was that forces like gravity or electromagnetism acted over arbitrary spatial distances without the need for any contact or physical exchange of any sort. For example, the sun keeps the earth revolving around it, just as the earth keeps the moon orbiting around it, kind of like twirling a ball tied to a string around our fingers. But out in space, there are no strings or any tangible signs of contact among the sun, the earth, and the moon. The force of gravity, which keeps them interacting and bound together, seems to reach out across space, invisible and intangible. It is a similar scenario with the electric force that keeps electrons bound to the nucleus in an atom or with magnets pulling little iron pieces from afar—they also seem to act mysteriously at a distance. To be fair, this spooky action at a distance was never satisfactory to Newton, even after he came up with the idea of universal gravitation that accurately described the motion of everything on earth and in the heavens. In fact, it bothered him so much that he did not publish this *magnum opus* of his life for a very long time, partly because he did not have an answer to this most puzzling question. Newton pushed the limits of what was humanly possible in the seventeenth century, but that question was way ahead of its time, and the answer had to wait for over two centuries until Einstein's theory of general relativity and the development of quantum mechanics clarified those issues. Einstein recounts his first encounter with a compass as a defining moment in his childhood that drew him toward science; he was absolutely fascinated by the mysterious force of magnetism that was invisible and intangible and yet powerful enough to keep the compass needle always pointing in the same direction, no matter where it is. I wonder if he had some premonitions about his defining role in finding the secrets of such invisible interactions.

Unless you happen to be a professional physicist, you would not have had any occasion to bother about the hows and whys of the interactions in nature. So the chances are that if you were asked, "How do you think interactions happen?" you would assume that the question is about social interactions, and you might very well say something like, "Oh, you talk, you touch, or you exchange something, perhaps a gift or perhaps blows, depending on the nature of the interaction." You could add, "Social and personal interactions absolutely rely upon exchange and interchange among people, be it words, looks, hugs, kisses, touch, or even bodily fluids." These are answers that make perfect common sense. But with these answers, you would have hit the scientific nail right on the head and explained how interactions happen in the quantum world, as well!

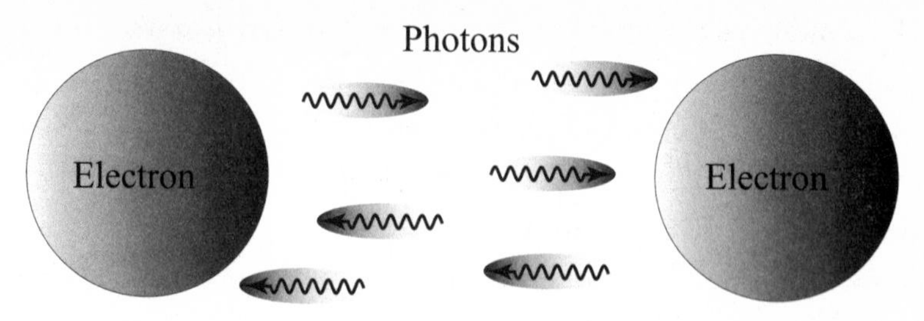

Figure12.1 Two electrons interact to repel each other by exchanging photons, or particles of light, which mediate the electromagnetic force.

You see, the quantum view of interactions is all about exchange: Interactions are mediated by the exchange of certain interaction-mediating "messenger" particles. Thus, the electromagnetic interaction that acts between electrons and protons holding atoms together and makes magnets work is due to the continuous exchange of photons or particles of light, as depicted in Figure 12.1. Gravity is thought to be mediated by an elusive particle named the *graviton*. The weak nuclear interaction associated with radioactivity is mediated by three elementary particles called W^+, W^-, and Z bosons. The strong nuclear interaction, which holds quarks together to form protons and neutrons, is mediated by particles called *gluons* (and appropriately so, because Superglue has nothing on them, because we can never pry two quarks apart, as we saw in Chapter 10). There is one caveat, though, that I should mention: Gravity continues to be a bit of an exception, because we still do not have a quantum theory of gravity, and that is one of the big outstanding challenges in physics today, so the graviton picture is not a done deal, and gravity is still best explained by objects moving along their natural trajectories in curved space-time according to Einstein's theory of general relativity, as discussed in Chapter 10. But for all the other fundamental forces of interaction for which a quantum understanding exists, interactions happen through exchange or some mediating messenger particles. Yes, light is the messenger that makes electricity and magnetism work—isn't that something! After all, as we saw in Chapter 1, light in turn is a manifestation of electromagnetic interaction at work.

This idea of interactions happening by exchange of some messenger particles is intuitive. It is certainly a much more comfortable and familiar notion than the classical notion of action at a distance—familiar, because human interactions operate exactly the same way. People do not interact magically with other people at a distance without some form of contact or communication. As illustrated in Figure 12.2, for people to interact, there needs to be physical, verbal, visual, or auditory contact or exchange: your eyes meet, you exchange a look (involving a whole bunch of photons); you talk to each other, you exchange words and sound; your lips meet, you exchange a kiss; you have direct physical contact, perhaps you exchanged a hug or pats on the back; you make love, well, there you are exchanging a lot of things—make up your own list! Perhaps these days you might be spending more time talking on the phone, texting, or chatting on the net; well then, besides words, written and spoken, you are exchanging electromagnetic impulses, as well, via wires and satellites. So you see, with human interactions there

Figure 12.2 Just like in quantum interactions, people interact by exchange, as well—exchanges of handshakes, words, kisses and hugs, and so forth.

is never any mystery about how it happens; there is always either direct contact or some sort of exchange. Newton might have guessed the right answer if only he had looked toward human behavior for his insights instead of staying focused on the heavens, but in his defense, on gravity the jury is still out.

But for all that, we still tend to overlook the crucial importance of exchange in relationships, particularly in this day and age of uber-independence and all-about-me culture. And I confess, so did I—until the truth about interactions was brought home to me forcefully by the ashes of one surreal relationship a while back. It was based entirely on a mutually agreed-upon unspoken policy of "I won't ask if you don't"—so all we had were brief and intense interludes that we spent together. For sure, it was great for a while—I felt like a married bachelor—I had my cake and could eat it, too, for a while. But that relationship could not last, because the interactions were brief and fleeting with extended periods of absolutely no exchange at all in between. With no meaningful exchange, whether made up of words or expressions of interest in the other person's life, or hugs and kisses, or even silly fights, there can be no relationship; you might as well be alone. Like the forces of nature, it is that exchange that binds people together. The sanctimonious put-down, "You're behaving like a married couple," hurled by exasperated friends at an incessantly bickering pair, is actually truly and profoundly telling of relationships, because the bickering is quite often an integral part of the exchange that keeps married people together. Some of the most enduring marriages we see are the ones that involve harmless and well-intentioned bickering, usually a sign of a healthy interest in the other person's life.

Likewise, there is truth in the belief that siblings who fight a lot as kids end up being very close in later life. That is certainly because they have that early exchange to build upon; tussles of the past become shared memories of the future, while others might continue to drift apart in peaceful detachment forever. As with biological siblings, so with fraternities and sororities; despite all the negativity associated with it, hazing exists for a reason—it acts as a bond-building ritual. The exchange and interchange forced upon the participants lay the foundations for stronger future ties and bonds; many who have gone through it will testify that some of the strongest and most lasting friendships they formed were forged on the anvil of hazing. Of course, as with everything, this is also a matter of degree, and things can, and do, get out of hand sometimes.

But all interactions are not the same, and there is a broad division of nature's interactions based on the nature of the particles being exchanged: interactions are long range or short range depending on the mass[16] (or weight) of the messenger particles. Some of those messenger particles could be quite heavy (on the scale of subatomic particles), and others have no mass at all. The photon, for instance, is massless, and therefore always travels at the maximum speed possible in the vacuum of space—the speed of light. Because they are massless, they can in principle travel forever all the way to the ends of the universe, as borne out by the fact that we can see quasars and galaxies billions of light-years away. But the messenger particles for the weak nuclear interaction, the W and the Z bosons, are quite heavy by elementary particle measures, so the distance they can travel is extremely limited—it is tough to get heavy things to move around. This essential difference has profound implications for the nature of the electromagnetic force and the weak nuclear force. Because the photon is massless and has no

[16] See Chapter 11 for a more detailed explanation of mass and weight, which is traditionally defined as the amount of matter in any object. A generalized meaning of mass as a measure of inertia is also discussed in Chapter 13.

travel restrictions, so to say, the electromagnetic force can act over large distances, causing the "action at a distance" that puzzled scientists for so long. So, we see electromagnets in junkyards pulling up things from afar. The weak interaction, on the other hand, is significant only at very short range—about the span of atomic nuclei—and that is why it never manifests directly on the scale of everyday life. Following the same arguments, although the graviton has not been detected in experiments, it is expected to be massless.

The spatio-temporal range of human interactions can be similarly classified based on what is being exchanged. When people are exchanging only the sound of words or pixels on the screen, they can be far apart and still be interacting—both sound and images can be carried far and away by electromagnetic waves—so we can talk and chat with people on the other side of the world. If people want to get beyond that and get physical, they need to be very close, in contact really, and they would be exchanging much more substantial things than the intangibles. Interactions that involve exchange of the tangible, the material, or the physical—things of substance—usually require close proximity or contact; they are short-range interactions. And like the short-range interactions in nature, human relationships that are primarily based only upon the physical and the material are unlikely to survive separation in time and space.

Interactions can compete—in nature and among humans. For instance, even though gravity feels strong, particularly when you are trying to climb a long flight of stairs, it is actually quite weak compared to electromagnetism. So if you hold up an iron ball with a magnet, your little magnet is actually counteracting the gravitational pull of the entire earth! In that competition, the magnet won. The strength and range of interactions determine which one will dominate in any situation. Inside an atomic nucleus, the residual strong nuclear force holds a whole bunch of protons together even though the electromagnetic force among them wants to tear them apart since they are all positively charged. At longer distances, like outside an atomic nucleus, the electromagnetic force dominates. Similar situations play out all the time in life. A strong attractive interaction can keep people together. But a stronger interaction can break apart a weaker interaction. So, if the wife ran off with the notorious traveling salesman or the ilk, there must have been a stronger interaction there than the long-standing but perhaps steadily weakening interactions with the husband, quantified by declining exchange of words, hugs, kisses, and sexual contact. On the other hand, even if the spouse has an intense infatuation with a rock star or a matinee idol, long-standing and strong interactions at home will probably keep it together even through temporary onslaughts where the object of desire is in town posing for pictures and signing autographs for adoring crowds.

If most people occasionally indulge in a bit of self-interaction, there is no reason to feel guilty about that. It is rather *natural* in the most basic sense of the word: It so happens that self-interaction is an essential part of the plethora of quantum interactions that define the subatomic world—particles constantly emit and reabsorb quanta of interaction-mediating particles.

Interactions can significantly alter behavior. Group dynamics can be quite different from individual behavior; extremes can be seen in mobs where the collective interactions among the individuals can completely alter the behavior. As we will see in Chapter 14, resonant interactions can amplify suppressed violent tendencies in a mob where often good-natured and friendly people suddenly become supremely vicious, capable of perpetuating unbelievable violence and mayhem. The behavior-altering impact of interactions is simply an extension of the amazing changes that interactions can induce in the material world. For example, in ordinary conductors of electricity, like in the wiring around the house, there is always significant loss of energy because of the resistance offered by the wire (just like water flowing on level ground slows down and eventually stops). But there are special materials called *superconductors*, used in a lot of high-tech equipment, in which electricity can flow with almost zero resistance. As a result of superconductors, electric currents can continue to flow for an extremely long time. But the reason superconductors are different is because of the onset of extraordinary long-range interaction among pairs of electrons via mediating particles called *phonons*[17] forming what are called *Cooper pairs*. In the laconic words of the Nobel Prize winning physicist Philip W. Anderson, "More is different," and the primary reason is the presence of interactions.

[17] *Phonons* are quanta of vibrations, including sound, so named in analogy with photons, the quanta of light.

CHAPTER 13
SEXUALLY BROKEN SYMMETRY

Spontaneously broken symmetry is the mechanism that gives mass and substance to all matter in the universe, via the famous Higgs boson. Sexual differentiation that breaks the genetic symmetry of asexual reproduction is the prelude to all our biological and cultural creations in ways that have evocative analogies with the way broken symmetry forges all material creation.

The creation myth of Adam and Eve getting evicted from Eden after tasting a forbidden fruit that induced an acute awareness of their nakedness has the thinly veiled implication that their big mistake was the discovery of sex. The result was a permanent exile from a blissful state of eternal life to the travails of mortality. This story has certain metaphorical truth to it in the context of the evolutionary and biological destiny of our species. There is an even deeper parallel in the realm of physics, tied to the origin of all matter in the universe, related to the famous Higgs boson. Calling it the God Particle, although intended originally to underscore its importance and elusiveness, will seem ironically appropriate in the context of this new connection we will explore here, because the Higgs boson has a rather analogous role to that played by God in the primal eviction.

We certainly do not need mythology or science to convince us that sex is important. Our existence is dominated by it, and most people spend a lot more time thinking of it than they let on. In truth, much of what we define as culture and tradition derives directly or indirectly from one of three things: religion, our quest for food, and sex. The gradual disconnect of most people from agriculture and food production and the secularization of modern societies have diminished the cultural presence of the first two factors in recent years compared to, say, medieval times. On the other hand, with technology and media innovations, the cultural impact of sex and everything associated with it has skyrocketed—meaning courtship, dating, and mating rituals and all the mind and body games involved in the process. Songs, dances, art, fashion, movies, a large fraction of literature, both prose and poetry, revolve to a large degree around the drama of man and woman engaged in the infinite variations of the mating game. We might label the strong emotions involved love eternal and so forth, but the biological motivation is sex, no matter how unromantic that might sound. This is true everywhere in the world, from the simple rituals of primitive tribes to the multilayered traditions of the most complex societies.

In previous chapters, we talked about human relationships and about what attracts people to one another. But there is one important aspect of relationships that we have not touched on: the differences in perspectives and expectations that generally exist between men and women regarding relationships and sex. There is certainly some asymmetry there. This relates to attitudes and biological inclinations and is not about capabilities, since the last hundred years have established beyond doubt that given the right opportunities, both men and women are capable of similar things. But men and women are by no means the same, and the differences dominate the mating game and define a lot of our cultural references and innuendos.

The most obvious and critical difference is that almost any woman, in her prime reproductive years (there may be some exceptions, as with everything in life), can be more or less certain that she can get laid if and when she wants to, as long as she is willing to lower her standards sufficiently. But no man can ever be that certain, not even the most desirable and charming of men, unless he is willing to pay for it in some form, not necessarily monetary. We see this quite clearly in the animal kingdom among various species of mammals and birds, and you bet *homo sapiens* are no exceptions: the males have to compete and fight for the females; the female has the right of choice, and for obvious biological reasons. This fundamental asymmetry is the source of a lot of misunderstandings and miscommunications—because we tend to overlook that it is hard to overcome millions of years of evolutionary instincts with a few paltry millennia of societal rules.

Asymmetry, or broken symmetry, has important implications in our lives and actually even more profound consequences in nature. But asymmetry is inherently tied to the concept of symmetry itself, an equally important facet in life and in physics. We all have our instinctive idea about what symmetry means. After all, our notions of beauty and aesthetics are directly tied to it. Symmetrical objects are more pleasing to our eyes, and therefore art and architecture abound in geometrical figures like triangles and circles, since they have certain obvious symmetries, as shown in Figure 13.1. The human body is more or less bilaterally symmetric, like the triangle shown in the figure. Some studies have actually indicated that more symmetrical faces are generally considered more beautiful.

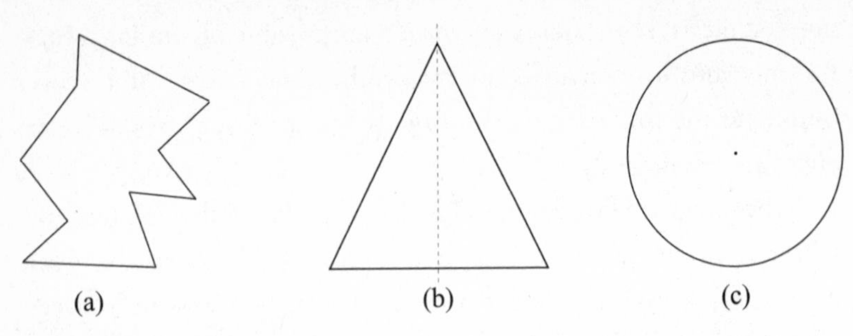

Figure 13.1 (a) This shape has no obvious symmetry. (b) This triangle has bilateral symmetry about the dotted line, meaning that its left and right sides look identical. (c) The circle has radial symmetry, since every point on the circle is exactly the same relative to the center.

But, beyond art and aesthetics, symmetry also has profound implications in science. In fact, a significant fraction of our physical understanding of the universe is based on the various symmetries that are preserved or lacking in the systems we study. The implications of symmetries are perhaps best contained in a fundamental mathematical law called Noether's theorem, named after the mathematician and physicist, Emmy Noether,[18] and it says, *With every continuous symmetry of a physical system, there exist certain corresponding quantities that are conserved, meaning that they do not change in time.* As we all know, in nature things are constantly changing, and one of our primary goals in our ongoing efforts to understand nature is to be able to predict how those changes occur. Therefore, it is of the deepest relevance that we identify those aspects that do remain unchanged as things change and evolve in time. In the jargon of physics, a physical quantity is said to be conserved if it stays constant as a system changes with

[18] Emmy Noether was one of the greatest mathematicians of all times. As one of the few women in the male-dominated mathematical world of the early twentieth century, she faced crippling sexual discrimination. As if that was not bad enough, being Jewish in Nazi Germany brought on persecutions that led her to eventually leave Europe and settle in the United States. It is a testament to her genius that despite such overwhelming odds, she made highly significant contributions to physics and mathematics.

time. Noether's theorem is extremely powerful because it allows us to determine those constants simply by looking at the symmetries possessed by a system! It would be no exaggeration to say that this theorem is one of the pillars of our understanding of the universe, particularly in our quest for the fundamental building blocks and the origins of the universe.

A few simple physical examples will give a flavor of how the theorem works. If a system has symmetry in regard to spatial translations, meaning that it does not change if we shift it to another position is space, then its momentum its conserved, which can be simply understood to mean that its velocity will not change (provided it does not fall apart and retains the same mass or amount of matter in it). Likewise, if a system is symmetric under translation in time, effectively meaning that it does not explicitly depend on time, then its total energy will remain constant and unchanging. Or if it is symmetric under change of orientation in space (like a circle, it still looks the same if we rotate it), then its angular or rotational momentum will be conserved, as in how fast it is rotating remains the same. These are just some of the innumerable applications of Noether's theorem.

It is rather interesting that we can apply the essential idea of Noether's theorem to certain aspects of our lives, as well. Let's consider what we started this chapter with: sexual asymmetry. In humans (as with essentially all sexually reproducing creatures), there is an inherent asymmetry in the sexual differentiation between male and female, each possessing certain quite distinguishable features. But not all creatures reproduce sexually; in fact, life had already been flourishing on this planet for about two and a half billion years before sexual reproduction came on the scene about a billion years ago. Even now, there are numerous creatures that reproduce asexually and do not have sexual dimorphism (that is to say two physically distinct sexes in the same species). So, from the perspective of physics, we could consider asexual creatures as being symmetrical (not in the geometrical sense), whereas sexually reproducing creatures have broken that symmetry by splitting into two distinct sexes.

Now if there is a symmetry, then following the idea of Noether's theorem we might ask, is there something conserved (that remains unchanging) in asexual reproduction and not in sexual reproduction? Absolutely! But it is certainly not the number of living entities. In both sexual and asexual reproduction, the numbers can multiply—a few parents can lead to many descendants. Instead, what is conserved is the genetic material. In asexual reproduction, the parent passes on an identical copy of its own genes to its progenies,[19] whereas in sexual reproduction, one half of genetic material is inherited from each parent, in effect creating new genetic material. Breaking the sexual symmetry leads to the violation of genetic conservation, meaning that genetic material gets significantly altered. It is rather amusing to note that even among sexual creatures, there might be some conservation rules if symmetry is imposed! Gay sex is symmetric as regards the absence of sexual dimorphism of the participants, and gay sex does conserve numbers—two people have sex, and at the end of that there are still two people. Straight sex between a man and a woman is asymmetric, and the conservation of numbers is frequently violated—at the end of about nine months, a completely new person can come into existence, sometimes even two or three or more!

There is actually a much deeper implication of this breaking of sexual symmetry. We will find intriguing parallels in some of the most esoteric ideas in physics that are tied to the ultimate

[19] There is also the possibility of changes to the genetic material due to mutations, but it is a slow process, which is part of the reason sexually reproducing creatures have been dominating evolution since sex came on the scene.

building blocks of the universe and the forces that define it. It is fair to mention here that the ideas in the next few pages might seem somewhat hard to grasp, but that is because these are indeed among the most complicated concepts in quantum physics, part of what is called *quantum field theory*. We will only touch on some of its general ideas here. Our understanding of the universe at its most fundamental level[20] defined by quantum field theory is based on two key insights:

1. Space-time[21] (even when seemingly empty) is permeated by certain physical quantities called *fields*, which can vary from point to point, meaning that they have "infinite degrees of freedom," since there are an infinite number of points in space-time. Electromagnetism (associated with electricity and magnetism) and gravity are examples of fields. We experience gravity although we cannot see it, and it can vary in strength at different points; for example, when we move up from the earth's surface, gravity gets weaker. It may help to think of fields as sort of like the Force in *Star Wars*—they can be intangible and invisible and still permeate space-time and have significant influence on all that exist, but, unlike in the movie, fields have no dependence on life.

2. All elementary particles—basic building blocks of the universe—are simply manifestations of changes or disturbances (called *excitations*) of certain fields. As astounding as this might sound, we have actually already brushed with the idea earlier when we mentioned in Chapter 12 that the electromagnetic force is associated with the photon (particle of light) and the weak nuclear force with two W and one Z bosons. We can now understand the photon

[20] Here we should understand "fundamental" in the sense of what has been experimentally established, which is at the level of elementary particles like electrons, protons, and quarks, and so forth—the ultimate building blocks we know of. There exist some very convincing and beautiful ideas about even smaller structures within some of those, such as described by string theory, but none of those have been validated by any experiments.

[21] Here we will frequently refer to space and time as a single continuum called space-time that can be visualized as a four-dimensional (4D) fabric or mesh.

as being an excitation of the electromagnetic field and W and Z bosons as that of the weak nuclear field.

The gravitational force has a definite direction at any point (we feel Earth's gravity pointing downward), whereas there are other fields that have no directionality. We can get an idea by thinking of the temperature (although not really a field) at different points in space; it has no preferred direction. The fields with directionality are called *vector* fields, like gravitational and electromagnetic fields, and those without are called *scalar* fields. At every point in space-time, there are multiple fields of both types present simultaneously, and

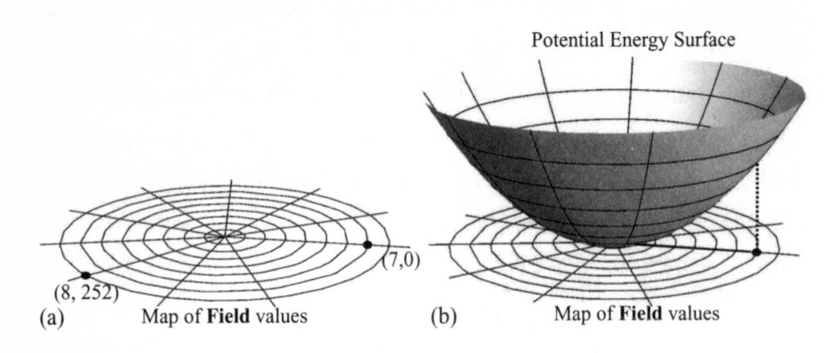

Figure 13.2 (a) The possible values a field can take may be represented by the position of points on a flat, circular grid, just like the position of a city or a town can be marked on a map. The field values can be uniquely marked with two numbers that denote its position on the grid like latitude and longitude in maps—two examples (black dots) are shown. The first of each number-pair represents distance from the center (count the rings), so that it would be zero at the center and increases outward: the farther out from the center, the larger and stronger is the field. (b) The parabolic surface maps the potential energy corresponding to the field values. For example, the position of the black dot marks a particular value of the field, while the vertical height (marked by the dotted line) to the parabolic surface measures the potential energy associated with that field value at the position of the dot. For this particular surface, the potential energy is at its minimum at the center where the field has zero value, but as we increase the field (move outward from the center) the potential energy increases as indicated by the upward sloping of the parabolic surface.

their interplay has profound implications for the existence of the universe as we know it.

Considering that gravity is an example of a field, it should come as no surprise that fields have potential energy (which is just stored energy as we saw in Chapter 4). At any point in space-time, a field can take any one of a continuum of values, just the way the temperature can vary. But as the field changes, its potential energy usually changes, as well—and how exactly it changes is of critical importance. An example is shown in Figure 13.2, where positions on the flat, circular grid define a map of values the field can take at *any one particular point* in space-time, and the height of the parabolic surface measures the corresponding potential energy. In this example, the potential energy is minimum when the field is zero (at the center). As the field increases (moves out from the center) the potential energy increases as indicated by the upward curving of the surface. Since the universe prefers lower potential energy, as we saw in Chapter 5, the field is most likely to have the value of zero. This becomes clear if we think of a ball rolling about on the potential energy surface: Its horizontal position marks the value of the field, and how high up on the curved surface determines its potential energy. Such a hypothetical ball will clearly settle at the center illustrating the preference for zero value for the field.

That was an example. In reality, all of space-time is infused with a scalar field that has a somewhat more complicated potential energy surface, in the shape of a Mexican hat or sombrero as shown in Figure 13.3. (We left out the flat grid that marks the field, with the understanding that the field is zero at the center and increases outward as before.) Clearly, now the potential energy has a local maximum (the peak of the hat) at the center where the field is zero. The minimum value occurs all around a ring (the valley of the hat)

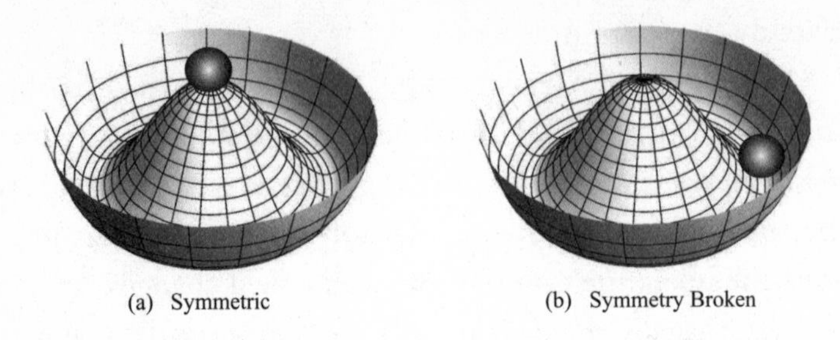

Figure 13.3 The potential energy surface in the shape of a Mexican hat is characteristic of a scalar field that permeates all of space-time. As in the previous figure, the actual value taken by the field is indicated by the horizontal position of a ball (although we leave out the flat, circular grid for clarity). (a) If the field happens to be at the center, its potential is a local maximum placing it at the peak of the hat shape, and its position maintains the radially-symmetrical shape of the potential energy surface. (b) If the field happens to have some value in the valley, then its lop-sided position spontaneously breaks that symmetry.

located a certain distance from the center. By the rule that lowest potential energy is preferred, the field is most likely to settle somewhere along that circular valley in the Mexican hat shape. Again, it is easy to visualize this by considering how a hypothetical ball on that surface would behave.

Notice that this potential energy surface has a nice radially symmetric shape about the center, and if the field (represented by the horizontal position of the ball shown in Figure 13.3(a)) has its value at the center, the scenario retains that symmetry. But when the field takes a value somewhere in the circle of minima (as in the ball lying at some point in the valley of the hat Figure 13.3(b)), the potential energy surface still remains symmetric, but the lop-sided positioning of the field (like the ball) on one specific side breaks that symmetry, as can be seen. The field is then said to have *spontaneously broken the symmetry* associated with the shape of the potential energy surface.

This *spontaneously broken symmetry* on this sombrero-shaped potential energy surface has certain deep implications as regards the origin of mass. But to appreciate that, we will need to remind ourselves of the broader meaning of "mass." In Chapter 11, we encountered the standard definition of mass as the amount of matter an object contains, so that heavier objects have more mass. But later in that same chapter, in the context of Newton's second law, we identified a more general meaning of mass as a measure of inertia of an object—a measure of its resistance to having its state of rest or motion altered by some external forces in the environment. Our everyday experiences confirm that: A massive tractor trailer is much more resistant to being moved about than a significantly less massive bicycle. "Massless" therefore simply means that the entity of interest has no such resistance or inertia.

Going back to Figure 13.3(b), we can see that if the scalar field tries to change or move in the radial direction, it will face resistance like a ball trying to roll up the steep sides. Therefore, such a radial "excitation" or disturbance of the field has mass in the sense we just described, and *so will the associated particle* (since field excitations are manifest as particles according to the second tenet of quantum field theory mentioned at the beginning of this discussion).However, if the field tries to change along the circular valley, there is no resistance since the potential energy does not curve up that way, so any excitation in that direction, and the associated particle, is massless. The massive particle associated with the field excitation in the radial direction is the famous Higgs boson!

This brings us to the heart of the matter: It so happens this scalar field is coupled to a vector field that also permeates spacetime—and the vector field is directly affected by the state of the scalar field, like people being affected by their spouses or partners they are coupled to. When the scalar field spontaneously breaks the symmetry and chooses an arbitrary point in the valley, then effectively we can ignore the possibilities of it being at any other point in that valley, so that we retain just that one slice of the full sombrero-shaped potential in Figure 13.3(b), along which the field (or the ball) is present, as shown in Figure 13.4(b). In the jargon of physics, this is called "fixing the gauge," meaning that we choose the one slice of interest from the infinitely many possible directions around the circle that the field could have picked (corresponding to the different points in the valley the ball could have rolled into).

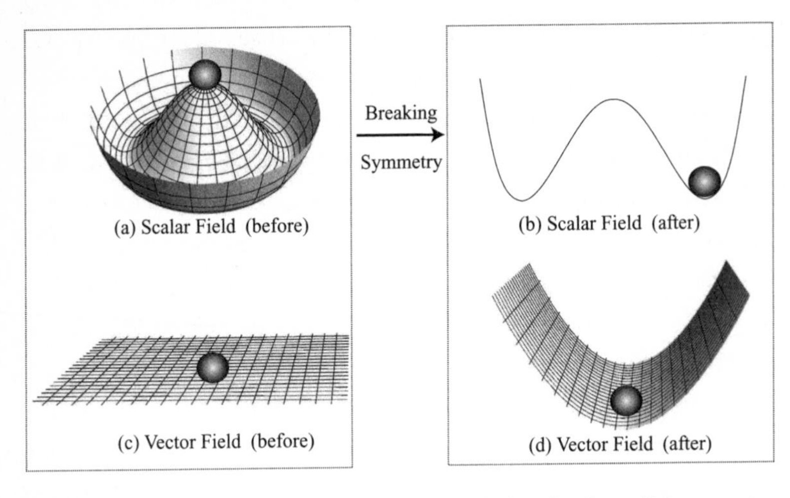

Figure 13.4 When the scalar field spontaneously breaks the radial symmetry present in (a), it chooses one point in the circular valley (represented by where the ball settles when it rolls off the tip). Thereafter, instead of the full sombrero-shape, we need consider only a slice of it, taken along the direction where the ball has settled, as shown in (b). This directly affects the vector field coupled to it. When the scalar field is symmetric, the vector field has a flat potential energy surface seen in (c), so the field (represented by a ball again) can change with no resistance, and the associated particles are massless. When the scalar field breaks the symmetry, (d) potential energy of the vector field rolls up in one direction, and the associated particles acquire mass since now there is resistance to field changes (like a ball rolling) in that direction.

But when this happens, the vector field changes, as well. When the scalar field is in the symmetric situation, the potential energy of the vector field is flat as shown in Figure 13.4(c) so that there is no resistance to changes in the field, meaning all particles associated with the vector field have *no mass*. When the scalar field breaks the symmetry, it makes the potential energy of the vector field roll up into a u-shape as shown in Figure 13.4(d). Field changes along the valley of the u-shape still face no resistance, so the associated particle remains massless. This particle happens to be the photon—the particle of light. But field changes up the steep walls of the valley face resistance, and the associated particle gains mass. This actually corresponds to three particles that happen to be two W bosons and the Z boson.[22]

This has profound implications, because this process known as the *Higgs mechanism* gives mass not only to the W and the Z bosons, but also to all the elementary particles like the electrons, protons, and neutrons that we are all made of—they would otherwise have remained as massless fields. Thus, the Higgs mechanism and the Higgs boson is truly and literally the origin of mass and, therefore, of all matter in the universe. Without it, the universe as we know it would not exist!

There is a very interesting analog to this in the evolution of life on this planet and how eternal life was lost. For the first two thirds of the billions of years life has existed on this planet, there was no sex—reproduction was asexual, whereby the offspring are exact genetic copies of the parent (via a few different mechanisms, for example, single-cell organisms might simply split in two). So these creatures never died naturally because their genetic code was

[22] Actually, the vector field has four directions, but that is impossible to draw, so to get the main idea across I have represented it as a flat surface (before symmetry breaking) that has only two independent directions. But in reality, when the symmetry is broken, the vector field "rolls up" in three directions (hence three massive particles) and remains flat in the fourth direction (corresponding to the massless photon).

preserved. On the other hand, offspring of sexually reproducing creatures are a blend of two distinct parents, so when the parents die, their specific genetic code is lost. Sexual reproduction has numerous evolutionary benefits due to the mixing of DNA from two separate parents, which can correct errors and also speed up advantageous evolutionary changes. However, it came at a price, because along with sex came death: Gradual decay and death are programmed into the cells of sexually reproducing organisms. A fascinating account of this can be found in a book with the evocative title, *Sex, and the Origins of Death*.

So sex and lifespan are indeed coupled together, just like the scalar and the vector fields we talked about. Think of sex as the analog of the scalar field and lifespan as that of the vector field—well, it is fitting in a way, since directionality has no meaning for sex, but it does for life, as it always moves forward in time. With no sexual dimorphism, asexual creatures maintain the sexual symmetry we mentioned in the context of Noether's theorem earlier. But there are infinitely many possible variations of how sexual dimorphisms appear (we just need to look at the different creatures on the planet or ponder other possible ways sexual reproduction might have happened). If we think of our own species, only one particular type of sex differentiation happened of those infinitely many possibilities—just like choosing one particular slice of the sexual landscape—and that breaks the sexual symmetry. This is figuratively sketched in Figure 13.5, which is quite analogous to Figure 13.4 but representing this new interpretation.

Viewing an organism's life-vector as the vector field, we can define eternal life as the lack of any natural resistance to living forever—just like a ball rolling on a flat potential surface represented in Figure 13.5(c). When sexual symmetry is broken, an absolute resistance emerges that prevents lifespans of sexually

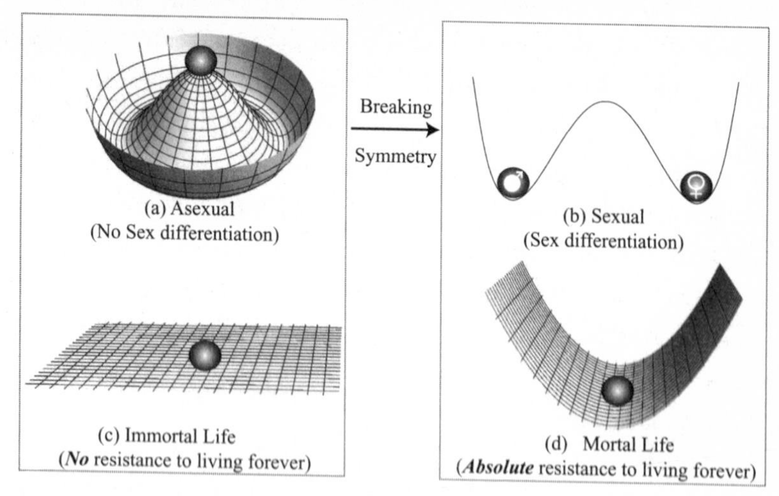

Figure 13.5 In analogy to the previous figure, asexuality has a certain symmetry figuratively represented in (a). When that symmetry is broken by sex differentiation, the specific sexual dimorphism for any organism picks out one possible slice, represented in (b), from the infinite variations among different sexually reproducing creatures. This fundamentally alters life-vectors of organisms to go from (c) no resistance to eternal life for asexually reproducing creatures to (d) absolute resistance to it for sexually reproducing creatures so that they are necessarily mortal with death inevitable.

reproducing creatures from "rolling on" forever. However, some creatures still remained asexual, so not all living organisms today are doomed to natural death; some immortals are still with us in the form of single-celled organisms and certain others—just as not all the vector field excitations acquired mass, only some did.

The analogy can be taken one fascinating step further, as we reflect on the consequences of having mass for elementary particles. You see, particles that are massive can never achieve the speed of light, according to Einstein's theory of relativity that we will see more of in Chapter 16. But, all massless particles *always* travel at the speed of light in vacuum—the maximum speed available. According to time dilation in relativity, time goes slower and slower as any entity approaches the speed of light so that time literally stands still at the speed of light. From that perspective, massless particles exist for an eternity. Massive particles exist in time—time can slow down, but it can never stand still, so they have lost eternity.

Thus, the Higgs boson associated with spontaneous symmetry breaking is instrumental in taking some of the fundamental particles necessary for the existence of the material universe and evicting them from the realm of massless eternity to be slow-moving particles dragged down by mass. So, perhaps in a metaphorical sense, the name of the God Particle is rather fitting.

CHAPTER 14
RESONANCE WILL SET YOU FREE

> *When the conditions are just right in the quantum world, things can zip right through barriers as if they did not exist at all! Magical as this might sound, the same principle—resonance—applies in our everyday world, as well. Harnessing that principle to work for us holds the key to effortless success, but getting on its wrong side could be catastrophic.*

The ability to move ethereally through solid walls and barriers is a must-have résumé item for any respectable witch or wizard. But actually, we do not need anything wizardry at all to send things though insurmountable barriers; it happens all the time in the quantum world, in a phenomenon called *tunneling*. The name "tunneling" is a bit of a misnomer, because no tunnels or holes-through-the-wall are involved, and the barrier remains absolutely unchanged during the process; things just end up on the other side like ghosts and wizards in movies. As a matter of fact, according to quantum mechanics, everything in the universe (including you and me) inherently has that possibility. The catch, however, is that (and as it so often happens with quantum phenomena) the probability of tunneling is significant only at very tiny scales—the scale of atoms and electrons. On the other hand, something as big as a person may have to wait a very long time—perhaps billions of years—to have any reasonable chance of spontaneously moving through walls. It is kind of like you have one in a billion chance of winning the Powerball, so you might have to buy a lottery ticket every day for perhaps millions of days to even have a reasonable chance of winning—but as with the lottery, the probability, although small, is never zero, and if you are lucky, you might spontaneously tunnel through in the first few attempts, but don't count on it!

So it appears that this quantum tunneling business does not seem very relevant for us in everyday life. Not directly, no—but there is a feature of this quantum tunneling that we can all harness in real life to breeze through obstacles in our way. It is called *resonance*. It is indeed a magical effect with far-reaching implications, and it contains the secret of navigating the hurdles of life effortlessly.

Think of a tiny subatomic particle—say an electron—hurtling toward a barrier with some speed, like a ball thrown at a wall. The ball would just bounce back, if the barrier were strong enough. And so would the electron—but only part of the time. Unlike the ball, even for a very strong barrier, some of the electrons can occasionally transmit straight through. In the jargon of quantum theory, there is non-zero probability of both transmission and reflection at the barrier. Sure it's a bit strange, but it is quite consistent with the whole probabilistic nature of quantum mechanics. But what if we could make the electron pass through the barrier 100 percent of the time, no matter how strong the barrier is? What if we could overcome even the limits of quantum probability and be absolutely certain that the electron would go through the barrier every time it is thrown at it? Actually, we can. Although this seems to defy both classical everyday logic as well as the probabilistic quantum logic, the magic of resonance makes it possible.

It has all got to do with the speed with which the electron approaches the barrier. If the electron comes in at just the right speed, it can go straight through the barrier as if the barrier did not exist! But the speed has to be just right. The analogous situation for the ball would go like this: We throw the ball at the wall at 10 mph, 20 mph, 30 mph, 60 mph, 100 mph, and it always bounces back, but if somewhere in between we throw it at 45 mph, it goes straight through as if the wall was not there—and it does so every time! It does not punch a hole in the wall, the wall stays intact; the ball just goes straight through. As we all know, that's quite unlikely to happen with a real ball—even once. But that's exactly what happens with electrons or other subatomic particles when we get the speed just right! Before you go off and think that it is about having a high velocity, let me assure you that is not the case at all, because if the velocity exceeds the right velocity, the electron would start bouncing back again. It is only at that one very particular velocity of approach, the barrier does not seem to exist at all for the electron. That velocity is the resonant velocity of approach, and the phenomenon is called *resonant transmission.*

You might think something like that could never happen in everyday life—but it does! Something very similar, following the same physical principle, is commonplace enough that you might very well have experienced it yourself. Have you ever been on a long straight city street with a series of traffic lights at every block? Every light is like a barrier that can make you stop if it turns red. But suppose there is very little traffic, and there are no cops around so that you can drive at your own chosen pace. Now you know quite well that if you drive at any random speed, you will soon hit a red light and will have to stop. But, if you drive at a particular speed that is synchronized to the time it takes the traffic lights to change, you could arrive at every light just as it is turning green. Then you would be driving through as if there were no traffic lights at all, as if you were on the freeway. Your driving has just become "resonant" with the frequency of the traffic light changes. The barriers posed by the traffic lights just disappeared for you, just like for a resonant electron—and just as with the electron, you know that you can't be going too slow or too fast; the speed has to be just right. In fact, traffic controllers often take the average speed of a car into account to adjust the timing of the traffic lights.

Unlike tunneling, resonance is not just a microscopic quantum phenomenon. It is all around us in the macroscopic world that we live in, and it is much more than an abstract idea—because every one of us is using resonance in some form or other, pretty much every day. In fact, we are absolutely dependent on it in this day and age. If you are fond of your smart phone, your television shows, Wi-Fi, and any of the wireless communication that none of us can seem to live without these days, you owe it all to resonance phenomena, because all of wireless communication is based upon the idea of resonance, and we will see just how in a bit. Resonance can also be a source of enormous power; it has been known to enable even a moderate wind to bring down a gigantic bridge.

Resonance applies to all arenas of life. It is no coincidence that often when we are in perfect agreement and harmony with someone, something, someplace, or even some idea, we speak of a sense of resonance. With an understanding of resonance and accompanying phenomena, we can really fine-tune our approach to people and to life situations to tackle hurdles and obstacles with a minimum of effort.

Resonance is directly associated with waves or periodic behavior. So its association with quantum tunneling is not a surprise, for after all in the quantum world, due to the wave-particle duality, everything has wavelike properties, so here we simply need to think of the electron as a wave. But waves transcend quantum mechanics, waves are all around us, and wherever there are waves or any periodic repeating behavior (like the traffic lights that turn on and off on a precise cycle), resonance can happen. Resonance is definitely one of the most useful things about waves and also the most striking. Yet the origin of resonance is simple to understand, particularly with the knowledge of waves that we already have from Chapter 9.

As we know from our earlier discussion of waves, the defining feature of any wave or oscillation is its frequency, which is how often it repeats itself. We can set any object vibrating or oscillating back and forth at any frequency we like; take, for example, a yo-yo on an elastic string—we could bounce it back and forth fast or slow, and the faster it is, the higher the frequency, because it completes each round trip more "frequently." Resonance happens due to the interplay of two quite different types of frequencies: one is called the *driving frequency*, and the other, the *natural frequency*. The driving frequency happens to be the one with no restrictions and has broader relevance in life as well as in nature. So let us talk about that first.

I bet even if you are not a scientist, you already know something about driving frequencies—because at some point in your life you most certainly have pushed someone riding a swing—perhaps one of your childhood friends or your own kid. The driving frequency here would be how frequently you push. Those pushes "drive" the motion of the swing, hence the name "driving" frequency. And it means exactly the same thing in any oscillatory system, mechanical or electronic, where an external force (like your push) drives the oscillations of the system (like the swing). The main difference is that in electronic systems, the oscillations would be invisible because they are oscillations of alternating current[23] flow moving back and forth in a circuit. Simple as it sounds, the electronic driving frequencies propel all of our wireless technology where all sorts of electromagnetic waves propagating through the air (transmitted via towers and satellites to our TVs, radios, and cell phones) act as driving forces in the various electronic devices and give us the all-powerful "signal."

Obviously, we can drive a system at any frequency we like; there is no real restriction—except that we don't drive it so hard that it breaks! That brings us to the idea of natural frequency. It so happens that pretty much everything that can oscillate has one or more natural frequencies at which it prefers to oscillate if it is left to itself. This can be compared to the harmonics or the stationary states on a plucked guitar string that we saw in Chapter 1. Unlike driving frequencies, which are external in origin, the natural frequencies are characteristic of the oscillating object itself and are restricted to certain specific values.

Resonance happens when we drive a system at one of its natural frequencies—meaning that the driving frequency matches a natural frequency of the system. The effect is quite dramatic: Since the driving frequency matches a natural frequency, the energy transfer from the driving force to the system is essentially perfect, because the system naturally wants to move at that frequency. The result is that the amplitude (or the size) of the oscillation increases with each push and can become extremely large, limited only by the damping or inherent resistance of the oscillating system. What is

[23] Electric current is of two types: direct current (DC) or alternating current (AC). The current from a battery is DC; it always flows in one direction from the positive terminal to the negative (see Figure 5.2). The current from the power grid, that we use about the house, is AC, which as the name suggests "alternates" or changes direction forward and backward very rapidly, like water sloshing back and forth. The motivation for using AC is that there is a very little loss of power over long distances, unlike for DC. Of course, as we see here, the oscillatory nature of AC is also absolutely essential for wireless technology.

quite amazing is that we somehow seem to know about resonance by instinct even as a kid. When we had to get the swing to go higher and faster without help from anyone, we all got to know the trick that it works best if we pull in our legs and push them out in sync with the motion of the swing. In doing that, we are essentially matching the driving frequency of the pulling-in-and-pushing-out of our legs to the frequency of the motion of the swing. If we do it just right, we end up swinging high and fast with the least amount of effort.

So if you ever wondered how all this wireless stuff works, it is all about resonance, and here's how. These days, we are literally immersed in electromagnetic waves of all kinds that carry signals for TV, radio, cell phones, and the internet. So how does your specific device pick out the correct signal? By resonance. Take the radio for instance, each station is sending its signal at a particular frequency (the driving frequency—fixed for each station[24]), so when you tune your radio, you are adjusting the resonant frequency of the electronics inside to match the driving frequency of the station you would like to listen to. All the other signals floating out there around you get filtered out because they have very weak effect—only the resonant frequency gets amplified, allowing you to listen to the station of your choice. That is the same basic principle in all of wireless technology, but of course with layers and layers of engineering sophistication added on to improve, clarify, and enhance the signal.

That is the science of resonance, but it also works its magic in social and human dynamics in many ways. Resonance and all that is associated with it can explain a lot about us. But as we already saw, resonance is a kind of special phenomenon and requires a precise

[24] Frequency, for anything repeating in time, is measured in units of Hertz (in short, Hz), which correspond to a frequency or repetition of once per second, so when you tune your FM radio to, say, 97.5, it means 97.5 MHz or 97.5 Million Hertz, meaning the electromagnetic wave that contains the "signal" is oscillating 97.5 million times per second!

matching of driving frequencies to natural frequencies. Likewise, in life, as in nature, we are more likely to encounter *nonresonant* scenarios much more often—resonance is something we need to seek out and work for. Before we go looking for resonance, we need to identify some of those driving forces in our lives and figure out how their lack of resonance with us might be seriously affecting us.

We all have our natural frequencies and rhythms, which we would follow if left to ourselves. But society, the people around us, and the demands of life and career impose all kinds of driving frequencies on us, and most of the time they are callously indifferent to our natural frequencies—they are not resonant with us. Say you are not a morning person and feel most productive at night—but do they care at your job? Of course not! So what happens in these more commonplace and more typical nonresonant situations that are still driven by external forces? Well, what happens is exactly what happens in nature, as I now illustrate.

One of the common complaints many guys have, a complaint that they can never air out at home, is that their living space and even their habits get redefined completely after moving in with a girlfriend, fiancé, and even more so, after setting up home with the wife. (If you are the wife/live-in-girlfriend, don't bother asking him, because if he is smart, he will deny it categorically!) Anyway, let's examine the guy's perspective for a bit. There you are, a happy carefree dude, going through life at your own natural frequency, then you fall in love, and if that lasts and goes anywhere, it inevitably leads to living together. Pretty soon, this beloved roommate of yours starts to drive you at her own frequency. It will start with innocuous things, like new bathroom rules, don't leave your shoes and socks lying around your bed, more tofu in your diet, maybe a yoga class or two, but eventually it escalates until one day you realize that you have been domesticated, and your habits are no longer yours, they are hers. What has happened is certainly not resonance! But it is a phenomenon that happens often in nature—take a system that is oscillating at its own natural frequency, then start driving it with an external force at a different but stronger frequency, not even close to any of the natural frequencies; eventually the system will start oscillating at the driving frequency, and its own natural frequency will disappear as a transient. Coming back to the perspective of that guy who was successful in love, this is pretty much what happens—the wife's driving frequency takes over and overrides his natural frequencies completely! But, to be fair, it may just as often go the other way depending on whose natural frequency is dominant, because the natural frequency of one is the driving frequency for the other.

If that seemed a bit partisan and biased, take a more universal example. America has become a sleep-deprived nation; people talk about it all time. The demands of our jobs require us to be at work at a fixed time every day, usually much earlier than many of us would care to wake up if we had a choice. Then often you do not get home till late, and then what with family chores and kids, your bedtime gets pushed way beyond what your tired body demands. So there you have your natural frequency defined by the biological circadian rhythm, and then you have the overwhelming nonresonant driving frequencies of your career, job, and family life. So, as in nature, the driving frequency takes over. Quite the same thing happens if you push a swing strongly at your own frequency with complete disregard for the natural frequency. It will take you more effort, but eventually you will get the swing going at the frequency at which you are pushing.

Your likes and dislikes and your preferred natural rhythms in life define your natural frequencies. If you were left to yourself, those frequencies would play out unhindered and freely. For example, if you like to play loud music at 3 a.m., and you happen to be stationed all by yourself in an Antarctic outpost monitoring the fluctuations of cosmic rays, then you can keep playing all the music you want as loud as you want to. On the other hand, if you live in the suburbs and are married to a light sleeper, whose need for the daily eight hours is understandably the more dominant daily rhythm, then you are in the presence of a strong driving frequency that is definitely nonresonant with yours. So what happens? You will soon cut short your natural predilection for loud music at 3 a.m. and fall into the rhythm of quietly sleeping at the time, or at least trying to. The driving frequency has overwhelmed your natural frequency.

Now, let's see what happens if you encounter resonance! Take the previously mentioned situation and now imagine that if you lived on the outskirts of town, and you married another nocturnal music lover; what do you expect would happen? You would be playing louder music longer and longer, because now you would be in resonance, so that your natural tendencies will be amplified—just like they do in nature.

Our very lives are defined by the day–night cycle and other cycles of nature. Our bodies are attuned to them. The wisdom of allowing your body to follow the circadian cycle is all about attuning your body's natural frequencies to the driving frequency of the day–night cycle. When it is perfectly tuned, you are in resonance and harmony with nature and your body's internal chemistry, and it will certainly make you healthier (and who knows, perhaps wealthier and wiser, too, as the old proverb says, "Early to bed, and early to rise makes one healthy, wealthy, and wise"). If you have any doubts about frequencies of nature affecting our bodies, then ask yourself, is it a coincidence that the women's menstrual cycle is roughly equal to the lunar cycle? In fact, the very word "menstruation" originates with the word "moon." And what is really amazing is that studies have shown that when women live in close association with one another, their periods often start happening around the same time; it is called "synchronous menstruation." We can view it as some sort of reproductive resonance behavior.

Like all important and relevant things in life, resonance also has a dark side. Among gangs and mobs, for example, resonance can amplify all the wrong human traits. We touched upon this a bit in Chapter 7 and Chapter 12. Everyone has some elements of a violent streak in him or her. But for most people most of the time, no matter how mad we get, it is well within control. But, in a mob situation, or hanging out with the wrong crowd, with a lot of people feeling the same way, those hidden characteristics resonate, and by feeding off one another, they get amplified; then suddenly people are capable of doing crazy, violent things that they would never dream of doing by themselves. Because when we are alone, such tendencies are relatively weak and subdued. Mobs amplify those through resonant behavior. Mob behavior can therefore be viewed as a manifestation of collective resonant psychology.

Even in nonresonant situations, when we are egged on by people around us to do things that we would normally shy away from—when we feel peer pressure—we are being driven out of sync with our natural rhythms and tendencies. If those driving forces of peer pressure are strong enough, they can truly take over our behavior, and we find ourselves acting in ways unnatural to us.

The power of resonance can sometimes be very dangerous both in life and in nature. One famous incident happened in Tacoma Narrows in Washington state in 1940, when a large suspension bridge collapsed primarily due to resonance. The bridge had a tendency to vibrate and sway even in moderate winds of 10 to 20 miles per hour. One day, the wind happened to sway the bridge at one of its natural frequencies, and as a result, the bridge started swaying with higher and higher amplitude due to resonance, just like a swing as we discussed earlier. As the amplitude of the oscillations increased, it was too much for the tensile strength of the construction material, and the bridge eventually collapsed, even though the wind was only at about 40 mph.

Similarly, resonance can wreck havoc in human life. Mob behavior that we mentioned above is just one example. But even in small groups, driven and resonant behavior can lead to amplification of the subdued negative qualities of people. For couples, that could be a serious issue. If someone has just barely controlled his or her drinking problem and then hooks up with a confirmed alcoholic, they will hit resonance, and chances are that they will both start to drink even more in the company of each other. That is just an example, but it goes for any habit we can think of. Just try to quit smoking when you hang out with smokers, or dieting when your family refrigerator is loaded with desserts, or going to bed early when your spouse stays up late at night. Resonance amplifies even minor characteristics and can make it very hard to suppress them.

But it also goes the other way; if you hang out with good people, your better qualities will be amplified. In college, if you hang with students who are serious about their studies, you too will be eventually; if you like to volunteer with charitable organizations and your spouse does, too, you will probably do much more volunteering together then you did alone; if you are an outdoors person and so is your spouse, you can count on an adventurous life together.

There are people out there who can naturally tap into the resonant frequencies of people, individually or as group. They are the ones who become great orators and public figures. As we say, their words resonate with the people. Indeed they do, in every sense of the word "resonate."

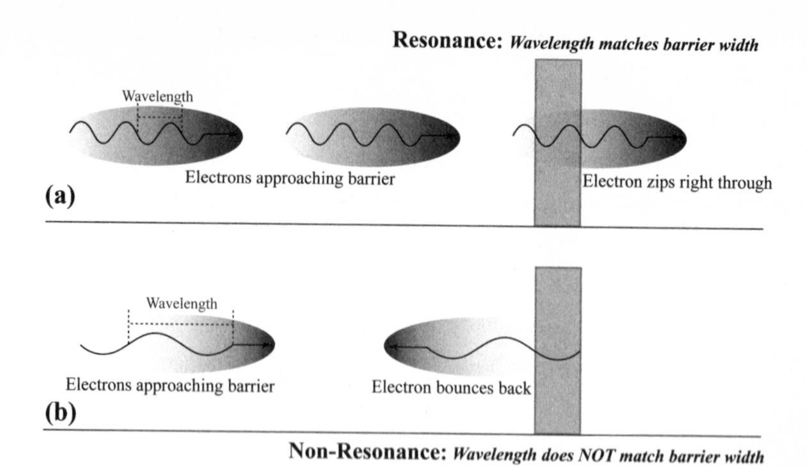

Figure 14.1 (a) At resonance, one or more complete wavelength exactly matches the width of the barrier, and the electrons pass right through. (b) There is no resonance if the wavelength does not match the width of the barrier, and there can be reflection of the electron. (Recall from Chapter 9, that the minimum distance or length over which the shape of a wave repeats is the wavelength of that wave.)

We will end the chapter by coming back to what we started with—how resonance allows an electron to tunnel through a barrier. In quantum mechanics, electrons can behave like a wave, and its speed is directly related to its wavelength (or equivalently its frequency)—the faster it moves, the shorter the wavelength of the electron wave. Now, if the electron moves at just the right speed so that its wavelength[25] exactly matches the width of the barrier as shown in Figure 14.1, then exact multiples of the wavelengths can "fit" within the width of the barrier, and then the electron-wave passes through the barrier as if it was not there. The frequency of the electron wave is its "natural" frequency, and the width of the barrier serves as the "driving" frequency (think of the two edges of the barrier "repeating" one edge after the other along the path of the

[25] Recall from Chapter 9 that the wavelength of a wave is the minimum distance over which it repeats itself.

electron). This is similar to the situation with the traffic lights—the distance (or "width") between the lights determines the driving frequency of how often the lights change colors, and the frequency of the driver crossing traffic lights is the natural frequency. When the driving frequency differs from the natural frequency, it forces an alteration of behavior—the electron bounces back, the car stops at a light. But when the driving frequency resonates with the natural frequency, all barriers seem to vanish away.

So all the wisdom about "being at the right place at the right time" and "striking while the iron is hot" and "timing your efforts for maximum return," and "buying low and selling high," and "tuning in to the pulse of the audience"—all of these are really about striking resonance.

CHAPTER 15
FAME AND THE EXISTENTIAL CRISIS OF SCHRÖDINGER'S CAT

The Schrödinger's cat phenomenon identifies the most bizarre aspect of quantum physics, in effect declaring that observation determines reality rather than vice versa, as we are prone to believe. Something that counter-intuitive is just what is needed to justify the inexplicable yet universal human need for companionship and craze for fame. Some ideas here may seem complex, but that's because they lead us to the subtlest depths of both quantum mechanics and human nature.

Robinson Crusoe is one of the best-known classics of English literature, an engrossing narrative of a castaway who survives and even thrives on a deserted tropical island paradise for almost three decades. But what is less well-known is that the novel was inspired by the real-life story of a Scottish sailor named Alexander Selkirk, who was marooned on an island and survived all by himself for four years until he was rescued. Both stories, the real one and the fictional, can be read as tales of survival, adventure, and human resourcefulness, but what really makes the stories timeless is their poignant and enduring message that none of us would ever choose to live alone forever even in a tropical paradise. Selkirk had made himself quite comfortable and secure on his island and could have chosen to continue as the "monarch of all he surveyed" (in the words W. Cowper's poem, "The Solitude of Alexander Selkirk"), but at the very first safe opportunity that presented itself, he left his island without a moment's hesitation.

We humans are indeed social creatures; we absolutely need to interact with other people. We start socializing from the moment we are born, making do at first with touching, gurgling, crying, and laughing—until we learn to talk, and then once we start talking, we just keep on talking and listening, and listening and talking, for the rest of our lives, some more than others, for sure. And we do pick up that first language with rare talent, don't we? A feat we can never, ever repeat again. Why? Because picking up that first language is absolutely necessary to start socializing and communicating seriously—it is never that urgent again!

This social nature is at the heart of being human—the very essence of our existence. To be loved, acknowledged, understood, and appreciated by friends and family, colleagues and acquaintances—that pretty much defines who we are and how we feel about ourselves. The people around us and our interactions with them indeed validate our existence, much more so than our pride and independence would allow us to admit. It is tempting to say that it is because of the convenience and security of living and working with others, but we know quite well that even with all that taken care of, we would still be left craving the society of other people—and that truly transcends anything tangible we desire. Therefore, simple and universal as this need is, it is hard to come up with a satisfying explanation for it.

That's because no one has ever thought to look in the realm of quantum physics. Incredibly enough, this quintessentially human need relates to one of the most bizarre aspects of quantum mechanics that goes by the cryptically innocuous name of *Schrödinger's cat.* But that's not all. This quantum connection also leads to a fascinating justification for some even more inexplicable aspects of human behavior, like our universal obsession with fame and our collective infatuation with celebrities. But perhaps the greatest advantage of establishing and understanding this unlikely connection is that it finally provides an intuitive feel for Schrödinger's cat itself, the strangest of all quantum phenomena—one that has puzzled generations of physicists, even the likes of Feynman and Einstein.

In the next few pages, we will encounter some ideas that might seem utterly unbelievable; therefore, it is worth prefacing them with a few comments. You see, no matter how crazy any of it might sound, all are well-established and experimentally verified consequences of quantum mechanics, which is the basis of much of our current understanding of the universe. In fact, quantum mechanics is the most successful understanding of the physical universe mankind has ever come up with; so successful in fact, that *everything* quantum mechanics has ever predicted has been verified to be true, even when those predictions totally defy common sense. Whether we hate it or love it, whether we are believers or skeptics, it does not matter; quantum mechanics has always been proven to be right. Einstein had a lifelong issue with it and tried hard to take it down, but in the end, quantum mechanics won and proved him wrong.

Before we tackle the really bizarre stuff, let us prep ourselves with a quick background on quantum mechanics. As we saw in Chapter 2, the *uncertainty principle* sets insurmountable limits to our knowledge, leaving plenty of room for chance in the universe. Particularly at the level of subatomic particles, we can never exactly pinpoint something (say an electron) and know everything about it; we can only specify the probabilities of finding the electron here or there or with one velocity or another. And that is true of all its relevant characteristics. In quantum mechanics, the collection of all the defining characteristics of something is called its *state*. For example, if that something happens to be your bicycle, its state will include all the relevant information about the bike—from it being in your garage, to its tires being pumped enough, whether it is standing upright or lying on its side, whether the brakes work, if the chain has been greased, and everything else you can think about it. Every combination of possibilities for all these features defines a definite "state" of the bike.

Now, it can very well be that you do not know exactly all the relevant information about your bike because you have not used it since last summer. Well, then you could speak of probabilities: There is 50 percent chance that the bike is standing up, there is 20 percent chance that the tire pressure is full, a 30 percent chance that the chain is greased, and so on. In the quantum world, the collection of such probabilities is contained in something called the *wavefunction* of that object (in this case, your bike). So the wavefunction contains the information on the probabilities of all possible states a system could be in.

For something large and complicated like a bike with so many defining characteristics, there are of course a lot of different possible states corresponding to the many combinations available. But for something tiny and simple like an electron, you need only to specify a few things to completely identify its states, like its position or velocity. But whether it is an electron or a bicycle, no matter what we are looking at, the key idea is that in the quantum view of the world, everything you can know about a system or an object is contained in the wavefunction, which specifies the distribution of the probabilities for all of its defining features. Even the whole universe is supposed to have a wavefunction! Once you know the wavefunction of any system, then you essentially know all that you could possibly know about it. However, there is something really, really sinister about the quantum wavefunction and the states in quantum mechanics, something that even after you understand it, you might still be hard pressed to believe . . .

Among the strangest consequences of quantum mechanics is this phenomenon known as *Schrödinger's cat.* The name comes from a thought experiment concocted by one of the key architects of quantum mechanics, Erwin Schrödinger, to illustrate the central role of observation and measurement in the quantum worldview. Here's how the thought experiment goes. We place a cat in a closed room, with an ample supply of air, food, kitty litter, and everything else the cat might need to survive for a while. But this is a room with a dark secret that the cat is happily unaware of—hidden in the room is also a deadly canister of cyanide gas that is rigged to be opened by a little radioactive triggering device, which has a fifty-fifty chance of setting off at any given instant of time. So we cannot know when the cyanide canister would open or if it would ever open at all. But if it does open, the feisty live cat would very quickly be a stone-dead cat.

To keep it simple, let us ignore everything else about the cat, and say that it can only be in two states: *dead* or *alive.* Now if the room is completely sealed so that we have no way of knowing what is transpiring inside, then we have no way to tell if the cyanide canister opened or stayed closed. Therefore, we have no way to determine whether the cat is still alive. Now, this is where the quantum strangeness comes in. According to quantum mechanics, until someone opens up the room and checks, the cat remains in a so-called *superposition* state of 50 percent dead and 50 percent alive—the wavefunction of the cat is in a half-and-half mixture of its "dead state" and its "alive state." The cat is not dead *or* alive—the cat is *both* dead and alive!

But wait, it gets stranger still. Quantum mechanics also tells us that the cat will remain in the half-alive, half-dead state *only as long* as nobody observes it. But as soon as someone opens the room and makes an actual observation, the state of the cat undergoes what is called *wavefunction collapse* to one of its two possible states, and it becomes *either* dead *or* alive depending on the fate of the cyanide canister while the door was closed.

This is actually so contrary to what we expect, that almost everyone misunderstands it the first time around, wondering, "What's the big deal?" After all, it just seems to be saying that the cat is either dead or alive, and because we don't know, we cite 50 percent chance of it being alive and 50 percent chance of it being dead—quite simple. No, not really—because the quantum view is that the cat *actually* exists in a state of *both dead and alive* until we open the room and observe it!

Physicists are not in the habit of doing experiments with cats. However, at the microscopic level of atoms and subatomic particles, superposition states have been created and demonstrated. Schrödinger's thought experiment highlights the crucial point that superposition events at the microscopic level (like with the radioactive process) could have unusual implications (via intermediate mechanisms like the rigged cyanide canister) for the macroscopic world we live in (as in a half-dead, half-alive cat).

The paradigm-shifting implication of the Schrödinger's cat experiment is that the physical reality of an entity, even a mortal one like the cat, depends upon the act of observation (in this case, by whoever opens the door to the sealed room). Anything that has not been observed exists *simultaneously* in all possible states available to it. That is its reality. *It is not as if the object exists in a particular state and we just do not happen to know which one. It actually exists in all available states until observed!* The act of observing does not simply give the observer the knowledge of that state, the observation *determines* the state and defines the physical reality.

Half-dead, half-alive cats are not easy to visualize and even harder to draw! So let us consider a less extreme scenario—that of an electron confined in a box, as I show in Figure 15.1. Before we observe the electron, it is equally likely to be anywhere in the box. So if we divide up the length of the box into ten equal sections, there will be equal probability of 10 percent of it being in any one of those sections. Let's represent that in another picture just below it by drawing a bunch of little rectangles at each position, all with equal heights corresponding to a probability of 10 percent.

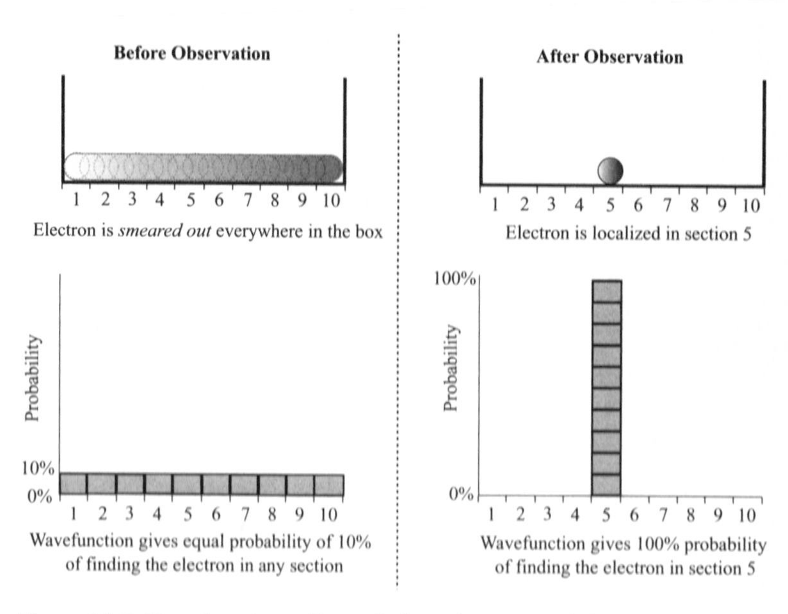

Figure 15.1 Wave function collapse: Before observation (left panels), the electron is everywhere in the box simultaneously as shown in the upper left panel, and therefore its wavefunction gives a flat probability shown in the lower left panel. After observation (right panels), the electron is found to be definitely in cell 5, and its wavefunction collapses to give 100 percent probability there and 0 percent elsewhere, as shown in the lower right panel.

Now to really appreciate why this is "quantum" strange, it is important to realize that the interpretation of "probability" here is not the same as with rolling dice in a Las Vegas casino. "Equal probability to be everywhere" in the quantum sense means the electron is *simultaneously* everywhere in that box—the electron is sort of smeared out uniformly over the entire length of the box! This is *completely different* from what we would normally understand by 10 percent probability of finding the electron in any section, which would be that the electron is sitting comfortably in one particular section, and we just do not know which one.

Suppose now we use a sophisticated microscope to observe the electron and happen to detect it in section 5, then instantaneously, the electron goes from being a smeared-out blob to being a nice, well-localized particle sitting there in section 5. Now the probability is 100 percent for section 5 and 0 percent in all other sections. We can represent that by stacking up rectangles there with a total height corresponding to the maximum probability of 100 percent and have no rectangles in any other section.

We can think of the rectangles as representing the wavefunction of the electron.[26] So before we make an observation, all the rectangles are the same, so the wavefunction is just flat, but as soon as we observe the electron, the wavefunction becomes very tall and sharply localized in just one region. It seems as if the little rectangles in the wavefunction sort of collapsed from a broad, flat distribution to stack up to form a very narrow and tall shape, as a result of the observation. That is why this is called "wavefunction collapse." The act of observing causes the probability density and the associated wavefunction to collapse from a shape that is *flat and spread out* to one that is *narrow and tall in a small region.*

[26] Actually, the probability corresponds to the absolute square of the wavefunction, but that is a detail that we will not bother with here.

Because this is such an onslaught on our accepted worldview, it is worth reiterating the key point: In our everyday common sense view, we could think, "Ah, the electron has always been in section 5, we just did not know it until we observed." But in the quantum view, the electron really is everywhere in the box simultaneously, and it is the act of observation that makes it settle and localize into section 5. That pretty much captures a big part of what is different and nonintuitive about quantum mechanics!

The strange and unavoidable implication, that observation determines the state of a system, jars with our common sense notions of objective reality—a reality that exists whether anyone is there to observe it or not. "Are you still there, when I am not looking at you?" savors a lot of an *Alice-in-Wonderland* kind of dreamy strangeness. Physicists are still grappling with the notion, decades after Schrödinger's pussycat first meowed. I have to confess that I worried about it, too. Long after I had convinced myself with the proofs and mathematical arguments, it still continued to jar me; that was until I started wondering if the notion was really so contrary to our life experiences. At some point, I realized that our entire conscious human existence is a validation of the cat's existential crisis. Perhaps physics students trying to tame Schrödinger's crazy feline might get some satisfaction realizing that there are strong parallels with how our social existence works. Let us see how . . .

You know it is a cliché for the lonely to say, "I feel like I do not exist." I have seen that in movies, I have read that line in books, and I have heard that in real life. Being completely alone in life, with no one to love or care can literally feel like being dead. Unless others take notice of our existence by being our friends, caring family members, or sympathetic and collegial colleagues, we can pretty much negate much of our existence as we know it and start to echo that line, "I feel like I do not exist." To really appreciate that, consider a more specific scenario. Have you been in a bar all alone standing by yourself, trying very hard to pretend that you are just fine with a glass of beer in your hand and smoking that cigarette that you do not really care for, but you smoke anyway because it gives you something to do as you keep your lonely vigil? No? Perhaps then you are way too cool for that. Well, in that case, you might have gotten an ego-boost from seeing guys like that around in a bar or a club. So what goes through a person's head in such a situation? What is the mental *state*?

If you were him, you would feel as if you are there, but you do not really exist there, because nobody cares whether you are there, so that you are essentially invisible. No one has noticed you (observed you); no one has measured you up for what you are as an individual. You are acutely aware of everything and everybody around you everywhere in the bar. Your mental state is delocalized over your surroundings—the entire bar; just like the wavefunction of the unobserved electron over the entire box. But wait! Someone just walked over to you, a shapely, well-formed specimen of a sex that you are interested in, and starts a conversation with you, buys you a drink, or lets you buy him or her one. Ah! Your mind immediately focuses on your new acquaintance near you, and you are no longer aware of what else is going on in the bar. That interest, and that acknowledgment of your presence, has suddenly focused your mind and body into the narrow space occupied by you and your observer. Your mental wavefunction has just collapsed to that point in time and space.

Hey, if the attention of one person, one single observation, can alter your state of mind so much, imagine what a multitude would do—hence our never-ending obsession with fame. Schrödinger's cat knows all about our universal hankering and desire for fame and can tell us why. Don't they say in show business, "Any publicity is good publicity?" When we are famous, our existence gets supervalidated by the millions of people who are aware of our existence, and as a result our self-assurance grows. Fame is like a zillion observations of the cat, with no doubt left about its state. Think of a rock star. There on stage, he is a god, everyone hanging on to his every word; he is supremely confident and self-assured. His existence and importance is validated in a very visible way. But the chances are if he were not a rock star, he might have been just another one of those guys sitting alone in a bar. Have you noticed how many of these rock stars seem shy and diffident in real life? The attention (the observations) they get converts their personal wavefunctions from something dull and deflated to something sharp and precise and self-assured. The bigger the fame, the bigger the adulation, the more confident is the state of mind. Quantum states sometimes work the same way: the more measurements we do, the more confidently determined the state becomes.

Nothing underscores the Schrödinger's cat situation better than the phenomenon of celebrities. Really, what is a celebrity after all? The concept of a celebrity exists as a state in the minds of all the observers. Take away all the observers—people who know and recognize the celebrities—and they simply become nobodies. Call them fans, call them the public, call them whatever you like, without the observers, celebrities do not exist. So is it any wonder that celebrities are taking over our lives these days, because with the boom in information technology, anyone can explode into the consciousness of a zillion observers around the world—just post a viral video on YouTube or convince a network to follow you around with a video camera as you do the most mundane things. You do not need any talent, achievements, or capabilities to speak of—all you need is the validation in the form of observation by interested viewers all around. Schrödinger's cat has gone wild!

But on a more sobering note, all of human greatness as well as all of human infamy, all those who are loved and respected and cited as role models, and all the ones whose notoriety we condemn, they are all really defined by our collective recognition as such. If their acts were known only to themselves, no other soul privy to them, the great would not be great, the notorious would not be notorious, they would just be lost in the mass of humanity like electrons in the universe that have not been observed. They only exist because of our collective recognition; take that away, and they might as well not exist.

It is not just how you feel about yourself or how others feel about you. Observation and the presence of others around us fundamentally change our behavior in every possible way. Who among us behaves and functions the same way in private as in public—of course no one! There is a natural state, call it our characteristic state (or *eigenstate,* as they say in quantum mechanics—homage to its German origins), that we operate in when we are alone with nobody around, say locked up in your room, house, or apartment. But introduce just one other person, an observer—spouse, parent, kid, friend, or neighbor—no matter how close and comfortable we might be with him or her, we are no longer in that eigenstate; our behavior adapts to the presence of the person. The less we know the person, the more of an external observer he or she is, the bigger the impact on our behavior pattern. And in public, in the presence of complete strangers, our state changes completely! A comic extreme of this universal human behaviorism is played out in the episodes of the classic British sitcom, *Keeping Up Appearances,* where every thought and act of the acutely class-conscious protagonist, Hyacinth, is defined by her aristocratic "observers," real or imagined.

Yet people continue to be mystified, surprised, enchanted, shocked, and disappointed to learn "the *true* nature of someone"!

Well, we shouldn't, and simply remember the primary lesson of Schrödinger's cat—a person under public scrutiny is in a different state from when he or she is hidden and unobserved in private life, and the state is even susceptible to who and how many are around to observe. It's a fundamental law, true for you and me and true for everyone we know. The truth is that there is no single "true nature" for anyone; it is all defined by observers around. Every one of us is really a *superposition* of different behavioral patterns, and which one is manifest at a given time is determined by who is there to observe. If you are ever surprised by hidden dimensions of someone's personality, it is just that you inadvertently had a glimpse of some other states of that person that are manifest with a different set of observers around that does not normally include you.

Not so long ago, I watched an inspiring movie about this young man who felt the true path to happiness and self-realization was to be alone far away from people and society, but at the end he realizes when it is too late that "Happiness is real only when it is shared." Exactly! No matter how romantic and alluring the notion might be to be a completely self-sufficient mountain man free from the constraints imposed by society, in practice, it can never be, because we cannot go far by denying our social nature; we can never be truly happy in isolation no matter how idyllic the surroundings and how magnificent the situation. That is why marooned in a tropical paradise, Robinson Crusoe spent the years yearning for human companionship, and his real-life inspiration Alexander Selkirk took the first chance he got to leave the island that sustained him quite well for four solitary years. So it is that there are millions living in poverty but relatively happy because they have friends and family to share their lives with, while many of the privileged ultrarich possessing everything in the world continue to make their shrinks rich by unloading the misery of their lives because they cannot find anyone else to take a sincere interest in them or their lives.

Schrödinger's cat could give us all a few pointers as we ponder the reality of our existence—is it all an illusion? Well, we need others to tell us that. If you feel a bit insecure, shy, or lonely once in a while, it is just Schrödinger's cat making its presence felt. It is alright to crave attention and friends and companions around you, because otherwise your existence is truly diminished. Without them you might not really exist, literally!

CHAPTER 16
MY TIME IS NOT YOUR TIME

Einstein's theory of special relativity established that time is relative and that absolute universal time does not exist. This notion is still counter-intuitive today as we rush about our daily lives, slaves to the clock. But in truth, we already instinctively know that our perception of time is anything but absolute, and exactly how the "relativity of time" plays out for us determines how we experience and perceive all our life experiences.

There is no absolute time. Time is relative. When the twenty-six-year-old Albert Einstein proclaimed this in 1905 in his *special theory of relativity*, everyone (who heard about it) was astounded. More than a century later, the truth of it has been established beyond doubt in countless experiments. Yet even today, few appreciate or even know that time is relative, and even those who do, consider it counter-intuitive—as going against common sense. But really that should not be because, as we will see here, our experience of time is anything but absolute, and in truth, the way we experience time, even in our ordinary life experiences, is more in *qualitative* agreement with the notions of special relativity than with the convenient fiction of universal time that we have all come to believe.

It is not hard to see why we cling to the notion of absolute time. After all, modern life in civilized societies hinges upon the recognition of a universal time that everyone agrees upon. If you are supposed to show up for work at 9 a.m., it is tacitly understood that it is 9 a.m. for you, 9 a.m. for your boss, and 9 a.m. for everybody else in your office. If you are late, you would not get very far trying to convince your boss that you only appear to be late but aren't really, because it is not yet 9 a.m. for you—time just has been going slower than usual at home. The chances are that you would get an earful about either getting a new watch or a new job.

"Be on time!" Most of us may feel that's all we need to know about time as we rush through our lives, slaves to that little phrase. We have to wake up at a certain time in the morning, get the kids to school, be at work at a specific time, have meetings, meet deadlines, pick the kids up after school, cook dinner, watch our favorite sitcoms, all at designated times that are agreed upon by society, co-workers, family, and friends. But it is all based upon one assumption that we take for granted—that the flow of time is the same for everyone, that there is an absolute universal time that exists independently of us and everything else in the universe. Thus, we talk of the river of time, evoking visions of time flowing on at the same unchanging pace for all of eternity, and we just happen to dance and drift on it for a while as our brief lifespans play out. There is indeed something comforting about absolute time, as the one thing we can all agree upon in a world full of disagreements and differences. But in truth, time is neither absolute nor universal. Time is way more subtle, always changing and shifting, and it is certainly worth taking some time to understand time itself a bit better.

Now, on being told that there is no universal time, it is natural to wonder, "Is this simply about having different time zones? They do have different times on the East Coast and the West Coast." That's true, but an hour is still an hour, no matter which time zone you are in. If you have a video-conference with someone in another time zone, and the conference lasts one hour by your reckoning, you would never think that it lasted only ten minutes for the folks at the other end. But in the realm of Einstein's relativity, such things could happen! What's more, time itself does not exist as some detached entity! When Einstein proved that the concept of an absolute universal time is just fiction, he also profoundly redefined the very meaning of time, as we will see later in this chapter. But before we confront the meaning of time, we need to settle this peculiar business of how relativity can make time intervals appear differently to different observers.

It has all got to do with motion and movement. Einstein showed that time goes slower for a person in motion from the perspective of someone at rest. Let me illustrate with a little "thought experiment," such as Einstein was wont to use. Say an express train with large transparent windows zooms by as you wait beneath a clock on a station platform, and high-speed cameras take *simultaneous* snapshots of the station clock and of a clock inside the train as it passes by. Assuming that the clocks are working well and synchronized shortly before, you would expect them to show exactly the same time. But if the train were to be traveling really, really fast—close to the speed of light—and if the cameras can somehow take pictures equally fast (well, this is a *thought* experiment!), you would see something really strange: the clock on the train would be running slower (it would show an earlier time than the platform clock). No, there is nothing wrong with either clock; it is just relativity at work! This effect is called *time dilation*, because from the perspective of someone on the platform, time measured by the clock inside the train does go slower—time stretches out and "dilates."

You might well ask, "How is it that we never perceive something like that in everyday life when we are riding a train, driving a car, or taking the bus?" We do not expect to routinely see our watch running behind the station clock after a bus ride or a train ride. The reason is quite simple—the effects of *time dilation* are impossibly minuscule to notice at the speeds we are used to. For those effects to become really noticeable, we need to be moving at speeds comparable to the speed of light—but light travels at 186,000 miles per second, yes that is per *second,* not per hour! Even driving a fast car at 100 miles per hour is just about 147 feet per second (or 45 meters per second), a far cry indeed from the speed of light; before the car covered a mile, light will have traveled about 270 times around the Earth. But the universe is full of material things that do move at speeds comparable to the speed of light, and we have created some of them even here on Earth. In the huge particle accelerators like the ones at CERN in Geneva or Fermilab in Chicago, subatomic particles can be accelerated to speeds very close to the speed of light—and there the effects of time dilation are dramatically manifest. You see, most of these subatomic particles are extremely short-lived; they appear and disappear in a matter of microseconds and nanoseconds.[27] But due to their extremely high speeds, time slows down for them, and their lifespans extend hundreds of times longer than if they were at rest. Physicists can see the direct effect of the time dilation in the markedly higher number of particles that make it to the detectors than would be possible with their resting lifespans—because the detectors are sometimes about

[27] A microsecond is one millionth of a second, and a nanosecond is one billionth of a second. So, for a particle that exists for one microsecond, one second is equivalent to a million times its natural lifespan.

a half mile away from the accelerator ring where the particles are created, and, even at close to light speed, it would take the particles much longer than their incredibly brief natural lifespans to reach the detector, and almost all would "die" and disappear along the way. But their very high speed slows down their relative time and stretches out their lifespans, so that in reality, most of them make it all the way to the detectors.

The astounding consequences of time dilation are perhaps best illustrated by another one of Einstein's famous "thought experiments." There are two identical twins; one of them goes off in a rocket traveling at speeds close to the speed of light, travels to a distant star, and arrives back on Earth a few years later (according to her), but on her return, she finds that her twin who stayed back on Earth is now old enough to pass for her mother. Strange as it seems, this follows from tested and proven equations of time dilation in the theory of relativity. An intuitive understanding of this is elusive even for many physicists. For sure, all physics majors in college learn how to use Einstein's equations for time dilation, but not all of them can honestly say that they have pondered it deeply. So if that's how things stand today, you can well imagine how crazy the whole idea must have seemed a hundred years ago when Einstein came up with it. Even to those few who understood what he meant by time being relative, it probably seemed less like real science, and more like a sequel to H. G. Wells's science fiction classic *The Time Machine* published just a few years before Einstein's paper on special relativity. As it turned out, Wells's time travel idea was not so far-fetched after all. You can always fast-forward into the future by traveling really fast, but the trouble is you can never go back in time. Time may be relative, but it does not go in reverse.

When I first learned about it, Einstein's *twin paradox* seemed completely fantastic. But the notion of time dilation itself does not seem so strange to me anymore—not because I have been familiar with the theory of relativity for a long time now, but because as I get older it has come home to me that this concept of time being relative is not so divorced from our human experience of time after all. Time for me is definitely going a lot faster these days than it used to when I was a teenager! And I know that I am not the only one; I have talked with people around me, young and old, and I have read about it, and I have come to the conclusion that it is a universal human experience. This feeling of life seeming to go by faster as we get older is something most people sense at some point in their lives. In a qualitative sense, our experience of time pretty much follows a rather *analogous* basic pattern as in special relativity, *except not when we go faster but as we get older*. You see, absolute time is just as much of a convenient fiction in our life experiences. But to really appreciate the analogies in the human perception of time, we first need to understand how time dilation happens in relativity.

Special relativity is based upon only two assumptions. First, there is the very obvious assumption that *the laws of physics are the same in all inertial frames of reference*, which is a fancy way of stating the fact that if you are sitting in a windowless, soundproof compartment moving at a uniform velocity (say in a train riding on perfectly smooth and straight tracks), then there is nothing you could possibly do that would tell you whether the compartment is moving or standing still. That is why even in a fast-moving train, in between stations you can pretty much carry on as if you are sitting on the platform. The only subtle part about the statement is the use of the word "inertial." This just means that the train cannot be accelerating—that is, speeding up, slowing down, or changing directions. You would certainly notice a difference then. For example, if the train were to speed up, you would be pushed back against your seat. If you momentarily forgot that you were inside a train, then you might think that some unknown force was pulling you into your seat, and the physics would seem different.

Then there is the not-so-obvious postulate that *the speed of light is the same for everyone no matter how fast anyone is moving.* Stated like that, it might seem quite ordinary and rather trivial. It is anything but that! Let me convince you that it is indeed an extraordinary assumption, in stark contrast to our normal, everyday understanding of motion. Say you are a careful driver these days, maybe because of some unforgiving recent speeding tickets, and you are prudently cruising along the road at the speed limit of 55 miles per hour. Then, all of a sudden, you are jerked out of your reverie by a sports car zooming by you in the next lane at 100 miles per hour, making you feel for a brief moment as if your car had stopped moving. You feel that way because from your perspective, the other car is moving forward at 100 – 55 = 45 miles per hour relative to you.

Now instead of a car, say it was a pulse (or short beam) of light zipping by you in the next lane. Well, in reality light moves at the incredible speed of 186,000 miles per second, but for the purpose of our little thought experiment, let us assume that the speed of light is only 100 miles per hour. Then according to the second postulate of relativity, that pulse of light would appear to move at exactly *the same speed* of 100 miles per hour relative to you, as well as relative to the stationary cop with his radar gun sitting on the side of the road. *The fact that you yourself are moving makes absolutely no difference.* In fact, even if you speeded up and hit 99.99 miles per hour (assuming that the cop was too engrossed with his box of donuts to give chase to you), even then the light pulse will still appear to be going at 100 miles per hour relative to you. It is not only that you cannot go faster than the speed of light (that's the cosmic speed limit!), but also no matter how fast you are going (below light speed, of course), light will always be moving away from you at exactly the same speed. If this seems very strange, that is because it *is* indeed very strange! We cannot draw upon any of our own personal experiences to understand this, because this is completely contrary to what we experience in everyday life—if we go faster and faster, we eventually expect to match the speed of anything and eventually exceed its speed and overtake it—not with light! I can only offer you an analogy from a work of fiction—Alice's dreamy adventures in *Through the Looking Glass*, where the Red Queen says to Alice, "Now, here, you see, it takes all the running you can do, to keep in the same place." Well, from the perspective of a beam of light, we can run, drive, or even fly in a spaceship as fast as we can, but we might as well be not moving at all. So, just like when we were children, and we would have accepted the Red Queen's words simply as a truth of the Looking Glass world, we will have to take this quirky feature of motion at high speeds as a truth of our universe.

All the strangeness of special relativity that makes it so non-intuitive stems from this single, strange assumption. But that assumption, no matter how strange it may seem, has been verified in countless experiments to be a fundamental law of the universe. This makes Einstein a genius like no other, because when all were guided by the common sense borne of everyday life, he had the boldness and insight to realize the truth of something like this, which was in complete contradiction to common sense. See, most of us even have trouble understanding it a century later!

Time dilation is a direct consequence of insisting that the speed of light stays the same in all inertial reference frames. It is easy to see how, with another quick little thought experiment. Again we will assume for simplicity that light moves quite slow, say at only about 100 miles per hour. Recall that the speed of an object is defined as the distance it covers per unit of time (like miles per hour). Suppose

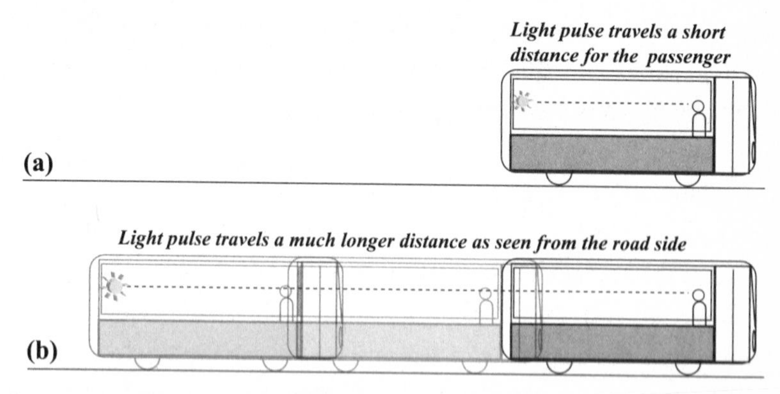

Figure 16.1 (a) As seen by a passenger sitting inside the moving bus, the flashing light source at the back of the bus remains at a fixed distance, so the distance traveled by each flash of light is just the length of the bus. (b) As seen by someone standing by the side of the road, each flash of light has to travel a lot farther to reach the passenger because while the light is in transit, the passenger would have moved forward.

you are at the front of a speeding bus, and a bright strobe light at the back of the bus flashes on and off briefly at regular time intervals. As seen by you, the light from each flash only has to traverse the length of the bus to reach you, as you can see in Figure 16.1(a). But as seen by the cop parked on the roadside, the light from the flash travels quite a bit farther, as you can see from Figure 16.1(b), because the front of the bus would have moved forward during the time the light is in transit. Now, here's the crux of the argument: *If the speed of light is the same for both you and the cop, but the distance traveled is different, then it must be that the time of travel of the light is also different for you and the cop.* To fully appreciate the point, imagine the strobe light is synchronized with a clock at the back of the bus to flash every two seconds—then your perception of two seconds sitting inside the bus, and hence the lapse of time measured by the strobe light, would be different from that of the cop looking at those same light flashes. Thus, insisting that the speed of light is the same for all observers regardless of their individual state of motion immediately leads to the unavoidable conclusion that the measure of time must be different for people moving relative to each other.

But hidden in this, there is another subtle yet more profound fact. Notice that the measure of time in our little thought experiment here was the interval between the two events—pulse emitted at the back of the bus *and* pulse reaching the front of the bus. Therein lies a universal truth that applies beyond the theory of relativity: Our perception of time really is determined by events. All clocks mark time by certain regular events—by a digit changing, or a needle or a hand sweeping across a dial, or a pendulum swinging. But, as we just saw, the time marked out can be altered even by changing the state of motion of those events. So, if we take away all events, what remains? Nothing, not even time! Thus, in a very literal sense, time does not really exist without some events to mark its passage.

If you could imagine the nightmare scenario of being locked away in an empty featureless room, where nothing ever changes, and say you do not even have bouts of hunger, thirst, and other bodily needs to mark the passage of time, then for you time would cease to exist. Everything that happens in our lives—waking up, breakfast, going to work, weekends, bedtime, vacations, moments of joy, sadness, hunger, thirst, and everything else we experience—give us the sense of passing time. But all those events are not the same in the way we perceive them. We naturally register unusual events more strongly than we do the mundane and the routine ones. You would certainly remember everything about your wedding day and the day your kid was born, but it would be quite unlikely that you would remember what you had for lunch a month ago, unless of course you have the same thing every day.

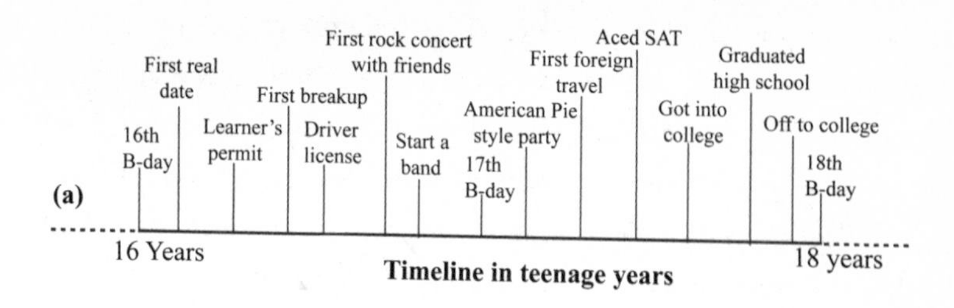

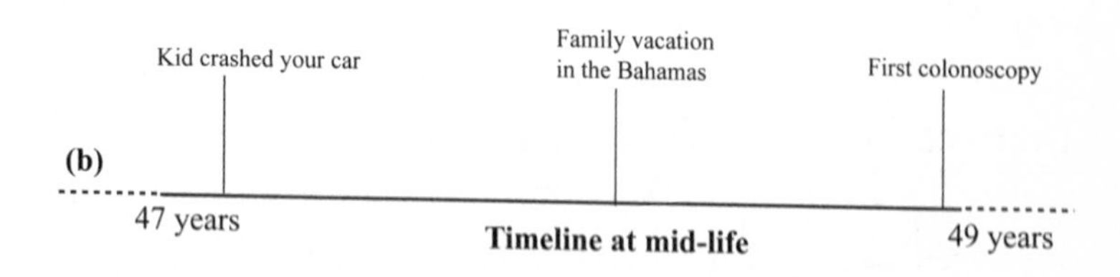

Figure 16.2 (a) When we are young, many memorable events mark the passage of time, so that the "unit of time = gap between such events" is smaller. (b) As we get older, fewer memorable events mark the passage of time, so that the "unit of time = gap between such events" is larger.

If our perceptions of different events are not the same, it stands to reason that our perception of the passage of time marked out by those events is not uniform either. When you are young, everything is new to you, so your mind registers lots of events consciously—the first bike, the first trip to the zoo, the first time your parents got really mad at you, the first day at school, the first date, the first kiss, the first breakup, and so on—a lot of firsts. All of those stick in your mind, creating a high temporal density of events with which to measure the passage of time. But, as you get older, you start to categorize a lot of things under "been there, done that," and you don't register many of those repeat experiences as strongly as you might have done in the past. So in your mind, there are fewer noticeable events to mark the passage of time.

Just for fun, I have drawn up an illustration in Figure 16.2 of what could be a typical timeline of memorable events over a two-year period between the ages of sixteen and eighteen years, and I compare it with a similar timeline for another two-year period later on in life between the ages of forty-seven and forty-nine years. *The intervals between those memorable events act as our true yardstick or "unit" for the measure of time*, because they are the ones that really mark our recollection of the passage of time. The two years between sixteen and eighteen are packed with events that stay with us, making the time "unit" smaller, so that we pass through several such "units" of time, and in our memories, it feels like that two-year period lasted forever. On the other hand, between forty-seven and forty-nine, only a couple of time "units" have come to pass, so these last two years seem to have just flown by in comparison.

Our experience of time is indeed relative and varies with age, sort of like how the passage of time varies with the speed in the theory of relativity. In relativity, as the speed increases, the unit of time seems to get stretched out compared to someone at rest, while in life, our perceived unit of time also gets stretched out as we get older compared to someone in their youth. After a couple of years of separation, a young man might say to his aging uncle, "Hey, long time no see!" and the old man might very well respond, "Really, has it been that long?" Well, that sounds a lot like what the twins in the twin paradox might say to each other after the trip to the stars.

Life is not being quite as objective as science, so one could take quite the opposite perspective, that time seems to go by quicker when we are having fun. Since we seem to have much more fun when we are young, shouldn't time seem to go slower when we are older and weighed down with responsibilities, and it is not so much fun anymore? There is some truth in that, and it is just another way to see how our experience of time is indeed relative. But this does not contradict what was said before: Sure, several days of vacation can seem to zip by much faster than a single hour waiting in line at the Department of Motor Vehicles (DMV), but that is only our perception of *time as we experience it at the moment,* which is quite different from how we recollect different periods of our lives and how they register in our memory. When we experience fun things, it might seem like time went by too fast because we wish they could have lasted longer, but when we remember those fun times, they do seem to stretch out because we remember so much about them. On the other hand, the hours we spent just waiting in lines or doing dull, routine things might have seemed never-ending at the time, but they just fade away quickly from our minds as if they had never happened.

As we get older, there are more of the DMV-like moments, of waiting and doing a lot of boring things that just have to be done. But those all slide by like a blur, and as we look back, years of doing such things seem to just roll by, and we do not even seem to notice—not at all like those nostalgic days when we were in school and the summers seemed to last forever. Our perception of time is indeed constantly changing. So, rather than be astounded by Einstein's statements about time dilation, we should instead wonder, how did we ever come to believe in a universal time? Well, you can't think about it for too long, or you might be late for your next appointment!

CHAPTER 17
OPTIMISM COLLAPSE ON THE WORLD LINE

Sooner or later, most people tend to look back and conclude that the optimistic outlook they might have started life with has faded with time; conventional wisdom would have us blame it all on the trials and tribulations of life. But that is not the whole story. The real reason can be defined by the notion of "world line" that tracks out the path in space-time for everything in the universe—a concept that originated in Einstein's theory of relativity.

"Ah! The optimism of youth," usually muttered with a wistful shake of the head and, of course, always by those who are not so young anymore—it is the nostalgic proclamation of our common belief that optimism and hope are at their peaks in our youth. And it is basically true: Sitting in a New York City subway, I used to look at the faces around me. Of course, you never look at anyone directly, that would violate the subway etiquette—you learn to look without seeming to look—I got quite good at that with practice, because I liked to study people's faces and expressions, trying to guess the story behind them. Anyway, what I wanted to say is that I was always amazed by the stark contrast that I invariably saw between the expressions on the youthful faces of teenagers and college kids—animated and full of life—and the vacant, empty expressions of the older crowd, mid-thirties and up on toward middle-age and beyond. In those days, I was still in graduate school, somewhere in between those two extremes, and I used to wonder what happens to people—and when does it happen—to cause such a visible and dramatic transition. Next time you are on any crowded means of public transportation, take a look and you might very well wonder, too. But I don't wonder anymore, because I have learned since then what happens: well, life happens.

If you are old enough, you, too, know that explanation quite well—chances are that you are living it right now—and you probably think it is because the young have not yet faced the harsh realities of human existence; the frustrations and disappointments of life have not yet had the chance to beat the hope and optimism out of them yet. True, that plays a role. But that is not really the main reason. To find the real source of youthful optimism and understand why it erodes throughout the years, it will help us to get familiar with certain evocative visions of space-time. Notions from quantum mechanics that we saw previously will also come in handy. But the key idea we need is one often associated with Einstein's theory of relativity—it is the concept of a *world line*.

The world line is a simple, yet elegant, concept—we just keep track of an object in space and in time, and the line that is traced out by its path would be its world line. Where it gets tricky is if we want to keep track of it in three dimensions of space plus one dimension of time, then we would have to visualize the path in four dimensions, and of course nobody can do that. But fortunately, that is a complication we do not need at all to understand the main idea. For instance, you can easily picture your personal world line by keeping

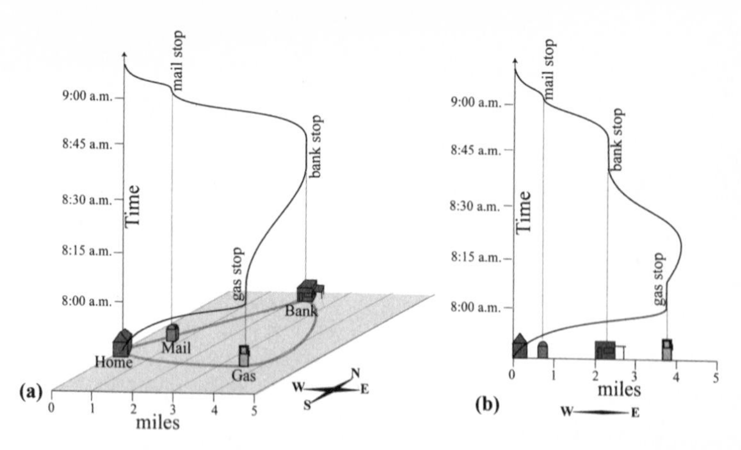

Figure 17.1 The thick line winding upward is an example of a world line for a person's morning errands run. The vertical axis always measures the lapse of time. The flat gray area at the bottom is a projection of our three-dimensional space on to two dimensions (like in a map) neglecting the up-down direction in space. (b) If we ignore the north-south directions as well, we can further project space down to just one dimension. Now, the horizontal axis represents only east-west movement, and the vertical axis keeps track of time as before. The world line in (b) is a projection of the world line in (a) on the east-west direction.

track of your movements, which usually occur on the surface of the earth and which therefore we can project on to a two-dimensional surface, like on a map laid out flat on a table, as shown in Figure 17.1(a). We can then measure out time upward in the direction perpendicular to the "map," so that with the two dimensions of space and the one dimension of time, there are only three dimensions in total, which is easy enough to draw and visualize. Now, if you did not move at all, your world line would be just a straight line going straight up because you would still be evolving through time, but if you move around, your world line will twist and turn, tracking your position as shown in Figure 17.1(a).

Now, there is a very important concept implicit in this picture. In drawing your path on the two-dimensional flat space like we do on a map, we are ignoring the ups and downs of the roads you might travel—meaning that we are ignoring the third (vertical) dimension. What we have done here is called a *projection* of a three-dimensional space to a two-dimensional space. We can go one step further and ignore the movement in the north-south direction as well and keep track of the changes in your position only along the east-west direction—then we would have a projection onto just one dimension of space, as you can see in Figure 17.1(b). The idea of projection can be generalized to any number of dimensions. So, while we might not be able to visualize say ten dimensions, we can always examine and draw their projections on any one particular dimension that we are interested in—and this notion will be very relevant to us here, as we now see.

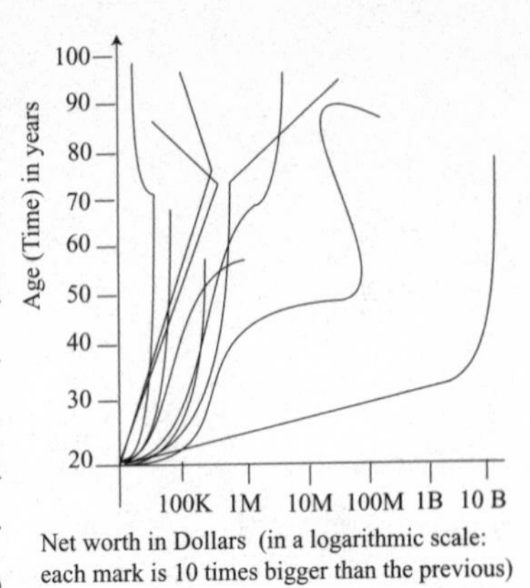

Figure 17.2 Possible projections of personal world lines on the dimension of wealth are shown. Such projections of world lines can be shown for every dimension of life. Each projection is like a shadow of the full multidimensional world line on that particular dimension of life, while we ignore all other dimensions.

In Figure 17.1, we assumed that we are only interested in your location in space-time. Well, keeping track of the location alone might work fine with things we usually encounter in physics, but people are a lot more than just inanimate objects. So to really describe how we change and evolve over time, we need to do more than just keep track of our location. This means that we need to generalize the concept of the world line by going seriously multidimensional, so that in addition to the usual physical dimensions that determine our position, we also add in additional dimensions to keep track of everything else about us (everything that we would care to keep track of)—such as our state of mind, wealth, job, relationships, spouse, children, feelings, emotions, and actions—over time. Each additional feature would add an extra dimension to the three dimensions of space and the one dimension of time we started with. That would be a pretty complicated trajectory, and we won't get far trying to visualize it in all its multidimensional glory! But it is not hard to get the general idea, and of course we can always look at particular projections corresponding to each dimension of our lives while we ignore all other dimensions, in exactly the same way we did in Figure 17.1(b), where we tracked the east-west dimension and ignored the north-south and up-down dimensions. In the same spirit, in Figure 17.2 we see some possible world lines showing only the projections that correspond to the "dimension" of wealth.

In Figure 17.2, we see a wide range of possibilities reflected in the financial dimensions of personal world lines, including being a multibillionaire. While the range of possibilities can be wide, it is important to keep in mind that not every possible world line is available at any particular point in space-time—in life as well as in the natural world. The physical limitations on world lines can help us appreciate that not all destinies are ever open to us.

World lines are often used in Einstein's theory of relativity to visualize the locus of an object or person in space-time. In such visualizations, illustrated in Figure 17.3, *light-cones* separate the possible world lines from the impossible ones; in Figure 17.1(a), space is projected on to a two-dimensional flat plane while the vertical direction tracks time. As mentioned before, the world line of a stationary object would be just another vertical line, since it evolves in time without changing position in space. But any sort of motion in any direction will tilt the line from the vertical indicating displacement in space—and the faster the motion, the

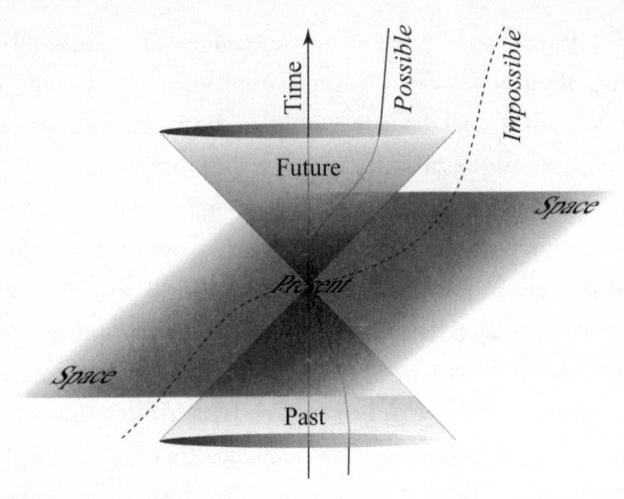

Figure 17.3 Space is projected on a flat plane, and time is measured along the vertical axis, just like in Figure 17.1(a). At each point in space at any given instant in time, like at the point marked as "Present," there are two light-cones—the past light-cone defined by all possible paths of light arriving at that point at that instant in time from all directions, and the future light-cone defined by the paths of light leaving it at that instant. Since nothing can move faster than the speed of light, all possible world lines passing through that point have to lie entirely inside the two light-cones. Any line that lies even partially outside them is forbidden by the theory of relativity, because they would correspond to motion faster than the speed of light. Note every point in space-time will have its own particular pair of light-cones.

larger the tilt. There is a *maximum tilt*, however, corresponding to the maximum speed possible in the universe—the speed of light, as we saw in Chapter 16. Thus, at any point in space-time, the possible paths of light, in all possible directions that pass through that point, mark out two cone-like shapes with their edges having the maximum tilt possible, as drawn in Figure 17.3—the *past light-cone* is defined by all the possible paths of light arriving at that point in space at that instant, and the *future light-cone* is defined by all possible paths of light leaving the point at that instant. Any world line that passes through that particular point in space-time *has to lie inside both the light-cones*; any line that lies outside, or even deviates outside, either of them even for a bit is strictly not allowed in the theory of relativity, because that would indicate that there is motion faster than the speed of light even if momentarily.

The implication is that there will always be fundamental limitations on our world lines, even when we apply it to other relevant aspects of our lives. For example, the chances of you or me ever becoming the ruler of some alien planet in a galaxy a million light years away are not just unlikely, but impossible within our lifespans. So, as we consider our personal world lines, we won't even consider such physically impossible ones. But even within the realm of allowed world lines and realistic terrestrial destinies, there are infinite possible ones, and we will talk about those.

Let us now figure out how your personal world line might evolve over the course of your life. At birth, your world line is not defined yet, so almost anything and everything (allowed by the limitations of nature and human capabilities and technology) is possible, and you might say that an infinite number of possible world lines are available to you—although most of them are of very low probability for sure. For example, if you were born a citizen of the United States, the probability of you being the president of the country someday might be very tiny, but never zero, as Barack Obama handily proved. (What probability would you have assigned to him at the time of his birth that he would become the president someday?) Borrowing from the idea of quantum states and superpositions from Chapter 15, we can say that at any given point in your life, you can view your future as a "superposition" of all the possible paths your world line could possibly take. At the moment of your birth, nothing has been defined, so technically all possible world lines are available, no matter how unlikely most of them are—unless of course you were born into utterly hopeless conditions, but in that case you probably would not have made it far enough in life to be reading this book anyway, so we will leave out that possibility!

So you start out in life with the maximal number of possible world lines open in front of you. Then what happens? As the years roll by, your life picks your real, actual world line out of the infinite possibilities, and your path gets more and more determined and specific, and your available choices diminish. With the passage of time, the superposition of the infinity of possible paths starts to collapse to the one path that you actually end up taking—for more and more of all those alternate possible paths that you did not take, the finite non-zero probability starts to collapse and disappear to become exactly zero. You can visualize yourself as walking through a labyrinth of tangled paths of possible world lines that stretch ahead of you. But as you walk on, the tangle of paths behind you disappears except for the one path you treaded on, and simultaneously, all those paths in front of you that were continuations of those vanishing paths behind you start to fade as well, leaving, with every step, a slightly diminished tangle of only those paths ahead that can be traced back directly to where you are now. And as more and more of those possible world lines disappear and fade into the nothingness of zero probability, so do many of your hopes and dreams that were built upon the unrealized promise of the open future of those vanishing world lines, which are now lost and gone perhaps forever. So is it any wonder that most people become less optimistic with time and age?

But surely infinite possible paths include just as many bad ones as good ones, so why be optimistic about having the choice? This is where the common wisdom about youthful optimism comes into play. As we anticipate the superposition of all those infinite possible realities that we call our future, with the heart and soul of youth, with little experience of failures and frustrations, we naturally expect life to follow the best and the brightest of those paths. That combination of not having tasted the bitter sides of life, and the choice of innumerable fantastic world lines still open to us, is what makes us so very optimistic when we start out in life. Then life goes on, and we make our choices (or are forced to make them by circumstances), and for most of us, the true world line we map out is almost never as satisfactory as the ones we had dreamed of—the glorious paths that we were so sure of, earlier on, start to fade away. In the language of quantum mechanics, the wavefunction of our life collapses bit by bit as the reality of life measures it out. Along with it, our optimism usually starts to collapse, as well.

Hope and optimism really are very much about having choices. At any given point in our lives, our level of optimism and hope is directly proportional to the number of viable options that are available to us at that time—the number density of realistically possible world lines. After all, what is the greatest complaint and source of frustrations of all who are dissatisfied and miserable about their lives? It is the feeling of being trapped in a particular life situation that is undesirable, tiresome, and frustrating. It could be a lousy job, a loveless marriage, bratty kids, an unsavory neighborhood, a tenacious disease. It is always the things that make us feel trapped that give us a pessimistic outlook on life. And that can only happen when our choices in life are very limited—after all, why would you continue in an undesirable situation if you had any real choice?

Of course, it is not just about the sheer number of options, but also about having the right kind of options—paths that are actually desirable. Just having a zillion world lines available to us at any time is not a guarantee for an optimistic state of mind. We should consider as viable options only those world lines that we would want to follow—world lines that lead us to a better situation than we are currently in. So we should qualify the earlier statement to say that optimism is really proportional to the number density of desirable world lines available to us at any given time. But of course, statistically, if there are lots of world lines available, it is quite likely that some of them will lead to a more desirable state. Whether we actually end up on the right one of course depends on our choices and circumstances. But just knowing that we have options available and paths open to us can change the state of our minds and give us hope and optimism. So here's a qualitative formula for optimism:

Optimism =

(density of desirable world lines available) × *(history of success so far)*

The first factor is what we just talked about. The second factor takes into account our usual wisdom about our life experiences taming our expectations over time—a history of success will make us view our options more optimistically.

By directly influencing our outlook on life, the evolution of our world lines explains much about our changing priorities with age. For instance, we can now easily understand our growing obsession with money as we get older. Kids do not care about money—at least, most normal kids don't! A business professor I once knew perhaps said it best, with a little twist on a well-known phrase: "Money cannot buy you happiness, but when you are unhappy, it sure as s— helps." That's because having money allows us to have more options, it opens up more world lines: If you hate your job, but you are in debt and living paycheck to paycheck, well, you are stuck with it. On the other hand, if you hate your job, and you just won the Powerball, then you have the option of quitting, because your windfall just opened up a whole bunch of very desirable world lines to lead you out of your work-related frustrations. Truly, the obsession of the adult population with having more money is mostly about increasing the available options, and opening up more possible world lines. Kids do not care about money because their options do not depend upon money—not yet.

Many wistful stories, poems, and movie plots have revolved around the "what ifs" of life—they are really speculations on our alternate world lines. Every moment of our lives is a "garden of forking paths" (the title of a fascinating story by Jorge Luis Borges); every choice we make in life picks out the path we take out of infinitely many. Even trivial choices can have profound impact. Say you stayed away from that party where you met your spouse, then you would not be married to him or her now, and the children you have now would not exist. Or say you decided to go back to your apartment one morning as you were walking to your car, to pick up something you had forgotten; those few minutes perhaps avoided a major accident that would have happened to you. But you will never know.

At least not in real life, but in the world of fiction you can. In a movie from not so long ago called *The Family Man*, the life of a man played by Nicolas Cage bifurcated into completely different paths as a result of one crucial decision, and we can see both of those realities. In his true reality—the world line his life actually followed—he is on the fantasy path, a successful Wall Street man, master-of-the-world type living large in Manhattan, but then through movie magic, where the laws of physics are casually suspended, he ends up on an alternate world line where he is just another ordinary Joe trying to struggle through middle-class life in the Jersey suburbs. Being a Hollywood movie, of course they had to try to get him the best of both realities in the end, but in the true reality of our lives where we cannot be so cavalier with the laws of physics, there is no peeking at our alternate world lines, and we definitely don't get to cherry-pick good bits out of multiple realities to patch together our ideal one. Wouldn't it be nice if we could! At least we can always dream and speculate and read books and watch movies.

Actually, the innate human interest in listening to stories, reading fiction, and watching movies originates in our deep desire to see and dream of other possible world lines. Stories, books, and movies all provide a window into the infinity of alternate world lines, which are unlikely ever to happen in the true reality of our lives. The stories that do come particularly close to our own reality or of someone we love and care about touch us deeply because they are world lines that we can identify with. Whether they touch us, fascinate us, or excite us, for that brief period during which we are immersed in the fictional reality, we are mentally traversing those alternate world lines. Often when we are very young, we tend to believe that some of those more exciting ones would actually come to pass in our lives. Reading *Treasure Island* or *Tom Sawyer* or *Huckleberry Finn* never again has the magic as they had when you read them as a kid, because then your world line still had not passed the ages of Jim, Tom, or Huck in those adventures, and the probability of you experiencing similar adventures is still non-zero.

We can meander and jump back and forth between multiple world lines as much as we like in literature and cinema, but in reality, jumping between world lines is not easy, if not impossible. With hard work and discipline and a bit of serendipity, we could certainly direct ourselves toward a more preferable but more difficult trajectory than the path of least resistance we naturally tend to follow. Many among us do succeed in doing just that, but the probability of success depends on a very crucial characteristic of world lines: They are very sensitive to initial conditions, meaning that even small choices and events can cause our world lines to diverge exponentially. For example, accepting a coupon from the airline to go on a later flight could save your life because the overbooked flight that you just opted out of is destined to crash with no survivors, or buying a lottery ticket on impulse, when you stopped to get coffee, could make you a millionaire overnight.

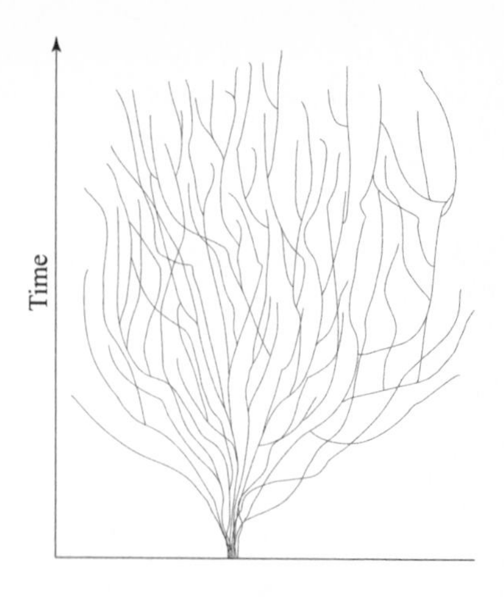

Figure 17.4 Visualizing how world lines diverge with time.

This divergence of world lines with time has important implications for our level of optimism as we get older. Whether we consciously acknowledge it, we are certainly aware of it in our cultural subconscious, as manifest in well-known aphorisms like, "It is not easy to change horses in midstream." Right, and with good reason, too! It is much easier to change course when you are starting out in life—to readjust your goals while still close to shore before starting to cross the turbulent sea of life. Soon after high school or a few years out of college, different life trajectories have not diverged much, as illustrated in Figure 17.4. The first couple of homecoming reunions reaffirm that—most of your fellow alumni are not that far off from you in their life situations; you can still connect with them and relate to them. With each subsequent homecoming, you are generally farther and farther apart—and everyone's life trajectory is a possible world line that might have been yours.

Changes are easy early on. As a recent graduate, if you are not happy with your career path at that point, you can change jobs; if you want to move to another city, you just pack up your stuff in a U-Haul and move out of your apartment; if you are having relationship issues, you just break up. Now, try doing all of that twenty years later: If you hate your career, too bad, you are stuck with it, it is next to impossible to seek out a new one now; if you hate the city you are in, well, your job, your house, and your family are all there, so you are stuck there; if you are having marital problems, well, now you need to go through legal channels for a divorce, and the prospects for a new relationship are not what they used to be, so you are forced to deal with it. You just can't jump to another world line so easily now, because the world lines are farther apart, and there are fewer available—there is real risk of falling into some very unattractive ones in between if you attempted such a jump at a later stage in your life, just as you risk going under if you try changing horses in midstream. There is so much more at stake.

From this perspective, it is easy to understand why as teenagers and twenty-somethings most of us are so cocky and so full of attitude—because we are absolutely certain at that point that one of those high-flying successful world lines we envision is bound to be our destiny, for after all everything is possible, and there is so much to choose from. With all those choices, our eyes are only for the high road, and it is easy to ignore the predominance of the middling or low roads all around. We are quite sure we won't turn into our parents, and we look with impatient bemusement at our older relatives and friends, wondering how and why they ever settled for so much less in life. Not for you the boring job, the nagging spouse, the little house, and the beat up old car. You will be driving an eye-catching import, living it up in some fancy place, earning millions in some position of authority. Without any serious failures to look back upon, as yet, you are certain of success where others might have failed. We all view our open world lines through the lens of our own hopes and dreams that blocks out the paths we would much rather not think of—call it the tunnel vision of youth!

Midlife crisis is just the onset of realization that you are on the wrong world line, and it is too late to jump to another; the world line you thought you would be on has seemed to have faded away completely, lost over the horizon. It is also the realization that changes are not so easy to make now. You might be able to buy the Porsche convertible if you are lucky enough, but your skin and your gray hair give it away as a prop for midlife crisis. If you are still single, you might dye your hair and try to play the dating field, but the chances are you will be seen as a potential sugar daddy or a cougar on the prowl. Well, the crisis blows over for most, and we learn to see and appreciate the good things in the world line we are on—once we accept that we are stuck on it, anyway.

It is best not to know of alternate possible world lines, but that knowledge is often forced upon us despite our best attempts. That is why reunions can be tough, and why we so often measure ourselves against how the friends of our youth are doing. Ever notice that there are lots of successful folks on the planet whose success never troubles us, but when we hear that someone we grew up with has made it big, has become much more successful than we ever will be—someone just like you and me, someone we always thought we were better than—that never fails to trigger a sense of "it could have been me," no matter how broadminded and altruistic we think we are. That is because they represent our very probable alternate world lines, because we shared very similar initial conditions for some time, but choices, circumstances, and fate took us on separate world lines.

So when they say, "the world is your oyster," to a youngster, they are really saying, "Hey kid, most of your world lines are still open." Act on it! If you are starting out in life, you should be aware that those world lines start closing fast, much faster than you might think, and small decisions you make now, that you might not think twice about, can have tremendous impact on which world line you end up on for the rest of your life. The lesson here is that the time to really engineer your fate is when you are young, that is, when you have the maximum choice of paths, and you can easily jump between them. With the right choices and some help from Lady Luck, your optimism just might last you a lifetime!

CHAPTER 18
IT IS ALL ON THE SURFACE

Surface features are often dismissed as being superficial. But certain crucial laws in physics tell us that all the relevant information of what's inside can frequently be gauged from the information on the surface—we just need to know what to look for. In life, actually, surface features are all that we have access to, although we may not perceive them as such.

"Beauty is only skin deep" is a rather scornful statement with a core of bitterness, uttered usually by people who do not feel so beautiful and, more often, by ones dumped and rejected by beautiful people. It advocates a distrust of surface features when it comes to gauging our fellow human beings—the general wisdom being that they could be quite deceptive about what lies inside. Perhaps there is some truth to that. But when it comes to people, the reality is that surface features are all we have to go by, even though we may not think of many of them as such. We can never read anyone's thoughts, so our perception of any person other than ourselves is always based upon what is on the surface. Of course, what is on the surface is much more than physical appearance; it includes what people say, how they dress and carry themselves, the level of confidence they project, their behavior, and everything else that we label in bulk as personality. But all of those qualities are as just as much on the surface as physical appearance, and we rely upon them to create our image of the person within.

There is absolutely nothing wrong with that, as is borne out by certain important laws of physics. There are plenty of physical scenarios where we can tell almost everything relevant about what lies inside from a knowledge of the properties at the surface!

One striking example from the physics of electromagnetism is known as *Gauss's law,* named after its discoverer, Carl Friedrich Gauss, one of the greatest mathematicians of all times. Gauss's law is an amazing and almost magical validation of the importance of surface features. This is how it works: Take any closed surface, real or imagined. Wait, first we need to clarify what is a closed surface. For example, the outside of a tin can along with its lid is a closed surface if the lid is on tight; on the other hand, if the lid is not on, the outside of the can is an open surface. Thus, a closed surface divides up space into two distinct regions—inside and outside—and there is no way to go from inside to outside or vice versa without cutting through that surface at some point; for instance, you can't go from outside a closed tin can to its inside without puncturing a hole in it somewhere. Now consider some such closed surface, which could be a closed tin can or just an imaginary surface that you visualize around any region of space. Now, suppose you would like to know how much *net* electric charge[28] is inside that closed surface—for example,

[28] Electric charge is just bits of electricity, the smallest bit being carried by the subatomic particles, electrons and protons. In a neutral atom, there are equal numbers of negatively charged electrons and positively charged protons; therefore, although everything is made of atoms, including ourselves, most objects do not appear "electrically charged." But sometimes, electrons can be knocked out of atoms, breaking the balance of charge inside atoms. This is what happens when winter coats rub against our clothes or body, and we build up "static electricity," meaning that electrons from the atoms in the coat are being knocked off onto our body or vice versa, making the one with a deficit of electrons "positively charged" and the other with an excess of electrons "negatively charged." Static discharge happens when we then touch another object, like a doorknob or another person, as electrons jump across to even out the imbalance.

if there are five units of negative charge and eight units of positive charge, the net electric charge inside would be 8 – 5 = 3 units of positive charge. The simplest way to find out would be to track down all the bits of charge inside, measure them, and add up their values. But that might not always be possible—for example, if the charges are in the interior of some extremely hot substance, or perhaps they are inside a very grouchy grizzly bear that swallowed a bit of electricity with its morning salmon. Then Gauss's law gives us an alternate way to accurately find how much charge is in a particular volume or region, solely from information on its enclosing surface or boundary!

Figure 18.1 Gauss's law states that the amount of electric charge enclosed by a surface can be obtained by simply measuring the outward force on a test charge at various sections of its surface, then multiplying those forces by the area of each corresponding section, and finally adding them all up. That information thereby gathered entirely from the surface features can tell us exactly how much charge is inside.

Positive and negative charge attract each other, while positive-positive and negative-negative repel each other. So, if we bring in a little bit of electric charge and put it anywhere on the surface, it will experience a force of attraction or repulsion. Now here's Gauss's magic: Divide up the surface into little sections, as shown Figure 18.1 (ideally, the sections should be as tiny as possible), and (i) measure the outward force felt in each section, (ii) multiply that force by the area of that particular section, and (iii) finally add up those products we get from *all* the sections. That will give us exactly how much net electric charge is inside the surface—without ever having to look inside the surface!

Not surprisingly, there is a very similar law involving mass and gravitational force; after all, both gravity and electromagnetism follow identical inverse-square laws, as we saw in Chapter 10. All matter in the universe exerts gravitational force on all other matter. Just as we did for electric charge, if we want to determine how much mass or matter is contained within a closed surface, we can figure it all out from the inward gravitational force felt at various points on the surface.

Thus, physical laws confirm that surface features do matter quite a lot and are not to be dismissed cavalierly. But to realize how in our lives we are completely dependent on surface features, we need to appreciate how truly inaccessible much of what is beneath the surface is. One of the most staggering thoughts we could ever have is to comprehend and hold in our minds, even if briefly, how utterly alone each one of us really is. It is not a thought that comes easily and is even harder to hold on to, because it is really like peering into an abyss. It is such an elusive thought that although it is easy enough to describe it in words, it is impossible to convey its dizzying feel unless you realize and feel it yourself—it is sort of like trying to get someone else to experience the grandeur of the Grand Canyon by describing it in words. But let me give the words a shot anyway: just try to realize in your mind that everything you have ever experienced, everything that exists, all of reality as you know it are all just in your head and in your head alone. No matter how close you are to anyone, you will always be trapped in your head and in your body, and you will never get to experience even the minutest thing from the consciousness of another person. You can never, ever know how it feels to experience life as another individual. Everything exists because you exist—as far as you really know. From your solitary perspective, everything will cease to exist when your consciousness ceases to exist. Our only validation to the contrary is that we see people around us die and disappear, and the universe continues to exist, and we extrapolate and assume the same has to be the case when we die . . . but we can never really be completely sure! As far as you are concerned, you are the most important person in the universe, and everything exists because you exist.

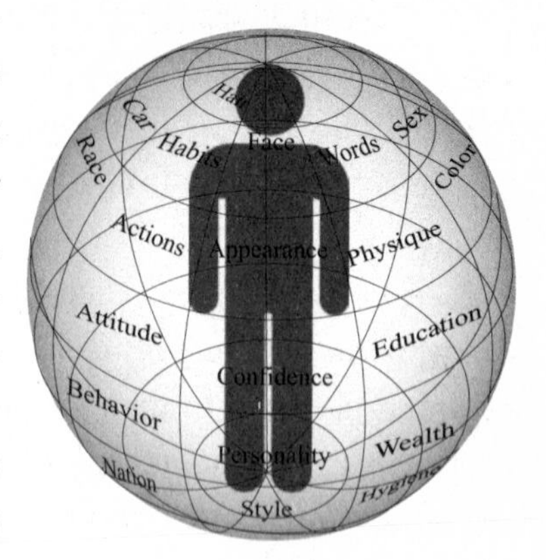

Figure 18.2 Just like Gauss's law, our knowledge of what is truly inside the mind and thoughts of a person is found by summing up all the surface information we can gather about the person. Since we never have all the information (hence the many empty grids on the surface), we will never know another person completely.

For sure, we all know that factually, but we seldom feel the full force of what it really means in regard to our conscious existence, which we take for granted. Once in a rare while, when the full implication of this touches my consciousness very briefly like a fleeting and vanishing remnant of a vivid dream, then there is that sensation of utter loneliness as well as an intense incredulity as to how among all the innumerable souls and consciousnesses possible, I happen to be experiencing this particular one and will only ever experience this particular one—and for that brief moment, it indeed feels absolutely dizzying.

Anyway, even if you never ever get that feeling and consider that entire thought utterly trivial, still the direct implication of this is transparent enough—we never really get to know the heart and soul of anyone else but our own selves, because we can never be inside another person's consciousness. So our image and judgment of everyone around us is in every sense based on what we see on the surface—their looks, their words, their actions, their choices and attitudes, and so on—as I show figuratively in Figure 18.2. We sum these all up to define the identity of the person inside those surface elements, very much like Gauss's law in Figure 18.1. But, even in the physical world, Gauss's law, although always true in principle, can be accurately used only when the surfaces are simple—like a sphere—where we can find the area of the surface easily, but if the surfaces are complicated, then doing the actual sum over the surface elements can be quite difficult, and only approximate answers can be found in regard to the amount of the net electric charge inside. People are even more complex than anything nature presents, and we will truly need to access an almost infinite number of facets to form a complete picture of any person. We cannot even come close to that possibility in our brief lifespans and the even briefer interludes of interaction with any particular person, no matter how close and intimate we might be. So our knowledge of anyone remains essentially incomplete.

The notion of all relevant information about a system being on its surface has found much deeper and more profound implications in recent decades. One of the most important developments in contemporary theoretical physics, simply known as the *holographic principle*, suggests (in its broadest and most speculative claims) that the entire universe can be thought of as being encoded on a two-dimensional "surface" that forms its cosmological boundary. The more rigorous versions of this essentially make a similar claim but not for the entire universe, rather for specific regions of space and all the matter and energy it contains—the description of such a region and basically all the information it contains is encoded on its boundary or surface. That volume of space might be the one that includes our solar system and all of humanity, as well. Unlike Gauss's law, the holographic principle[29] is not an experimentally proven fact, but rather an attractive and logically consistent hypothesis.

If indeed the surface information is so crucial, the natural question arises, how is it that we are so often misled by surface features, in judging people, nations, cultures, and situations? After all, aphorisms like "Beauty is only skin deep" and "Don't judge a book by its cover" have emerged from accumulated experience and wisdom of the ages. The reason is, although we may have access to the information on the surface, that information does not capture *all* the information, but only some essential aspects of it. That's true of analogous physical scenarios, too. For example, while Gauss's law can use the surface information to determine how much charge is inside the surface, it cannot tell us anything about how the enclosed electric charge is distributed and positioned inside the surface. This is illustrated in Figure 18.3—we can state with certainty that there are five units of electric charge inside the spherical surface shown. With Gauss's law, we would not be able to differentiate whether all of it is on one object at the center or distributed equally among five objects, and in the latter case, the five bits of charge could be located anywhere inside the surface without changing the conclusions of Gauss's law. There is a very similar situation with gravity: We can imagine an enormous imaginary surface enclosing the earth and the moon. By applying the gravitational equivalent of Gauss's law, we could tell what the total mass of the earth-moon system is by measuring the inward force at all points on the surface. But if the moon were to fall into the earth, we would still get the same answer, and if the

[29] The holographic principle is anchored in the physics subfield of string theory, which is based on the hypothesis that the fundamental entities in the universe are not point-like particles, but tiny "strings." The interest in string theory is largely motivated by the thought that it is one of the most promising routes to reconciling Einstein's theory of gravity with quantum mechanics, two great pillars of modern physics that have continued to be incompatible so far.

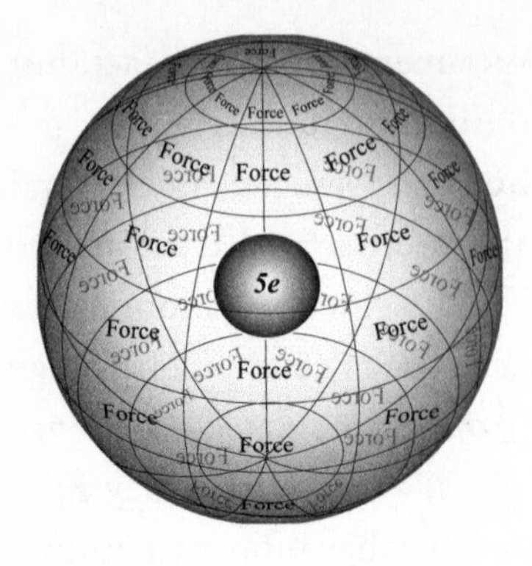

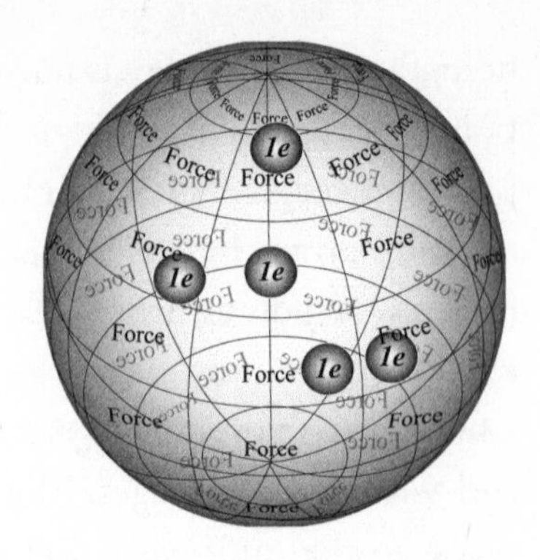

Figure 18.3 Gauss's law can tell us only how much electric charge is enclosed by a closed surface, but it cannot give any information regarding how that charge is distributed inside. So, for example, for both of the scenarios sketched above, Gauss's law would give the same answer because there are five units of charge (measured in units "e" for "electricity" or "electrons") in both cases. But obviously the charges are all in one chunk in one case and in five different chunks in the other. We could move the charges around inside the surface, but that also would not change the result of Gauss's law.

moon and the earth got compressed to the size of a peanut forming a black hole, we still would not be able to tell the difference from the surface measurements, because the amount of mass or matter within the surface remains the same.

This physical analog points to the root of inherent ambiguities in our perceptions of the society, people, and culture we live amidst. Much of the information that determines our decisions and opinions are based on surface features, and quite often that is all we have access to. Yet, as the physical example in the last paragraph indicated, while the surface contains a lot of information, it does not always contain *all* the information, so it can be manipulated and adjusted. Much of politics and marketing, particularly of consumer products, relies upon this flexibility of surface projections. The entire advertising industry is based on that, and all of propaganda past, present, and future hinges upon that—the reality and the truth are often deemed immaterial and cursory, while the surface image defines the perceived reality of our opinions and attitudes toward products and people, particularly famous people who we never get to know personally. Nowhere is this more rampant and utterly misleading than in the public perception of media personalities, particularly actors and musicians. The craze of the fans latches on to the image projected in movie roles and musical lyrics, and it often happens that even getting to know a few of the other facets of the real person is enough to disappoint even the most ardent of fans.

It is quite interesting that technology has slowly but steadily pushed us toward increasingly distilling the three-dimensional world we live in through two-dimensional surfaces: first innocuously through the printed word on the flat surfaces of newspapers and book pages; then through the movies and television screens; and then ultimately the granddaddy of them all, the computer screens and then tablets. Notice the TV and computer screens are getting flatter almost as if to stress the fact that they really are two dimensional in the end! While the innovations in technology are great things in themselves, it is disconcerting that much of innate fundamental structures of human interactions and human experience are being transformed in the process. In modern society, we seem to spend more and more time in front of the literal two-dimensional projections of the world and less time in direct interaction with people and in experiencing life in actual activities. The surface features are literally defining the very fabric of human existence today. Whereas in direct personal interaction, we may have some sense of probing what is below the surface, distilled through a computer or TV screen, what is on the surface is all that we have—and often a very fragmented and distorted version at that. The real issue here is that information distilled through the various media does not always provide comprehensive or correct information even about what is on the actual surface of events and the people around us.

The ancient fable about the blind men and an elephant succinctly summarizes the essence of how our perceptions of everything around us are to varying extents inaccurate because, even with what is available on the surface, we only have access to bits and fragments. The story goes that a group of blind men (or to be more inclusive, we can say people in a pitch-dark room) touch an elephant. One feels the leg and concludes that the elephant is shaped like a pillar, another grabs the trunk and concludes it is like a large snake, a third is swished on his face by the elephant's tail and proclaims the beast to be in the shape of a rope, a fourth feels a gently swaying ear and envisions a fan (not an electric fan but a traditional one), a fifth slides his hand up a tusk, and deems that the animal is shaped like a spear, and the sixth touches the body and thinks of a large box. So who is right? In the fable, they all argue and eventually get to the moral of the story—that they are all partially right, but no one is completely right because they each have only a fragment of the truth.

This goes to the heart of the matter, but not as the fable intended. The "truth" in this fable is simply the image of the elephant as a whole, but this still is just its surface and nothing else—not its biology, its nature, or its habits. Thus, without meaning to (and something that all the various versions of the fable fail to mention), the fable underscores that all the information we will ever have access to is on the surface, but what it does stress is exactly what is crucially important in Gauss's law as well—we need to have complete and accurate information of the entire surface, not just fragments of it. In life, we seldom have the complete picture; therefore, our knowledge of anybody or anything is inherently incomplete.

CHAPTER 19
THE EXCLUSION PRINCIPLE

Our need for privacy and our own space can be traced to certain eccentricities of the elementary particles we are made of. The exclusion principle classifies all particles into fermions, which won't share a quantum state with its kin, and bosons, which will share it with all. Judge for yourself whether you are fermionic or bosonic.

We cherish our privacy, and we highly value our personal space, and we can be very protective of both. At home, at work, and even in public places, we tend to feel most at ease if we can stake out our own bit of personal space—a private room at home, a cubicle in the office, a place to sit down in a packed room. Very crowded places make most of us feel uncomfortable; even people getting too close can ruffle us a bit. This was comically spotlighted in an episode of the sitcom *Seinfeld*, which featured the general irritation and distress caused by a "close-talker"—a fellow with the habit of standing unusually close to others when in conversation, in essence disrupting and invading their personal space.

This human preference for personal space and privacy is amusingly evident in public urinals, where most men predictably avoid using a stand right next to one that is occupied (assuming no other intentions are involved); a man, given the option, would typically keep at least one or two stands between him and the next guy. Other examples abound, and here's a non-gender–specific one: in a bus or a train, if there are plenty of seats available, people always make a beeline for the totally empty seats instead of sitting next to someone in a partially occupied seat. There may be exceptions, but these are always because of other overriding factors, like, for instance, a young fellow choosing to sit next to an attractive woman in a bus full of empty seats.

This obsession with privacy and personal space appears to be a very human trait. We would think at least it must be associated with life—after all, evolution, both biological and societal, and instincts of survival seem to have ingrained in us the need to look out for ourselves, including looking out for any bit of space we can call our own. Therefore, it might come as a big surprise to learn that this *exclusion principle* of keeping out others from our corner of the universe goes very deep—all the way down to the fundamental building blocks of matter!

In physics, we call those fundamental building blocks *elementary particles*; it will help to review some of their features. In earlier chapters, we have already seen that all matter is composed of tiny little atoms, and the atoms themselves are made of even tinier particles called protons, neutrons, and electrons. The protons and neutrons are tightly packed at the center of the atom, in a tiny clump called the nucleus, while the electrons revolve around the nucleus, sort of like planets around the sun. The protons and neutrons are of about the same size, but protons carry a positive electric charge, while the neutrons are neutral (hence their name). Electrons, on the other hand, are about two thousand times lighter in weight than the proton, but it packs the exact same amount of electric charge as the proton, except that the charge is negative. But, although so different in many ways, all three of these elementary particles—protons, neutrons, and electrons—share one thing in common: they are all *fermions*.

Fermions? What might they be? Well, answering that goes to the heart of this business of "quantum exclusion." All the elementary particles in the universe fall into two major categories called *fermions* or *bosons*, just like all living things are classified into plants or animals. This classification has to do with an intrinsic property of all elementary particles called the *spin*. We can think of it in pretty much the same way as we understand it in common usage: that the elementary particles like electrons are spinning like a top. But unlike a top, which can rotate at any speed, the spin (as in the rate of spinning) of an elementary particle is quantized, meaning that it can only take on special values, and here's the key point: the spins can take integer values 0, 1, 2, 3, . . . or half-integer values like ½, 3⁄2, 5⁄2, and so forth. *If the spin of a particle is an integer, then the particle is called a boson, and if it is a half-integer, it is called a fermion.*

Well, we classified the particles, so what? Actually it's a very big deal. This is not some simple bookkeeping trick like separating the yellow balls from the red balls. Bosons and fermions are different in some fundamental ways that underlie the very existence of the universe as we know it. It so happens that the particles with integer spin—the bosons—are a gregarious, friendly lot and love to hang out with their kin. The other ones with half-integer spins—the fermions—tend to avoid others just like themselves. And guess which ones we are all made of? Right—fermions! Protons, neutrons, and electrons all have a spin of ½, and they are most definitely fermions.

If we put two electrons that are exactly like each other in every way in close proximity, they will automatically prefer a configuration where they are farthest apart from each other. The reason I say "exactly like each other" is because there is an extra subtlety here: Although all electrons have spin ½, they still have the choice of spinning clockwise or counterclockwise, and that can make two electrons different from each other. In physics literature, counterclockwise spin is referred to as spin "up" and clockwise as spin "down," because if you curl the fingers of your right hand in the direction of the spin, then your thumb would point up or down respectively. So if two electrons are such that one is spin up and the other spin down, they tolerate each other's presence because they are not quite the same, but if both are spin up or both are spin down, they are exactly identical, and then they absolutely cannot be together in the same space! The probability of finding them in the same place becomes exactly zero.

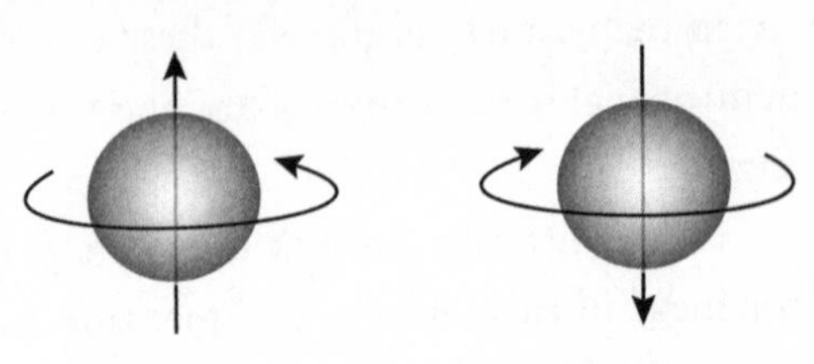

Figure 19.1 Counterclockwise spin is labeled spin "up," and clockwise spin is spin "down." That's because if we curl the fingers of our right hand along the direction of the spin, and we stick out our thumb, it will point up or down respectively.

We see rather similar behavior if we think of women and men as two groups of humans, just as spin up and spin down are two possible spin states of an electron. Traditionally, and in the overwhelming majority of cases even today, men and women eventually pair up and end up sharing their space and much of their private lives, like electrons of opposite spin. The stable pairing of men with men or women with women occurs less often, as a percentage of the population, just as electrons of like spin cannot share the same space. Moreover, since there are only two spins, adding in any more electrons—to make a triplet, quadruplet, and so forth—will lead to at least two of them having the same spin, so three or more electrons existing in the same place also has zero probability; well, that's sort of like a subatomic version of monogamy! But of course, with humans there are always exceptions to every rule, and this rule nevertheless gives us a curious analogy between the private life of electrons and the dominant trends in ours.

Well, we have talked about fermions, but what about bosons? Recently, the discovery of the Higgs boson at CERN made a splash in the news, so that it has become a familiar term. Much of the popular hype was generated by the boson's unfortunate nickname, the "God Particle." This particle has no more to do with God than any other particle out there in the universe. Nevertheless, the hype is well-deserved because of its role in giving mass to all particles as we saw in Chapter 13. The Higgs boson is a boson, with integer spin 0.

Well that is just one type of boson; are there others? In Chapter 12, we mentioned that the electromagnetic interaction is mediated by photons (particles of light) and the weak nuclear interaction by the W and the Z bosons—they are all bosons of spin 1. It is rather interesting that all these interaction-mediating particles are bosons and not fermions; if fermions are like people holding on selfishly to their corner of space, bosons are like the hugs and kisses, gifts, and interactions that can be stacked up, with no limits on their numbers, keeping the fermions together. Bosons are responsible for every physical interaction we know of, acting as the glue holding the universe together. That makes perfect everyday sense—you cannot expect an introverted individual averse to socializing to keep a party together.

But I should mention that the analogy only goes so far, because atoms themselves can be both bosonic or fermionic, even though the building blocks are all fermions, since the spins add up, and the combined half-integer spins of an electron and proton, for example, would add up to an integer spin. Nevertheless, it is rather interesting that the elementary particles we are made of are fermions with the inherent proclivity for keeping others of their kind out of their space. In fact, if you probe deeper down to the level of quarks that make up the protons and neutrons, they are fermions, as well!

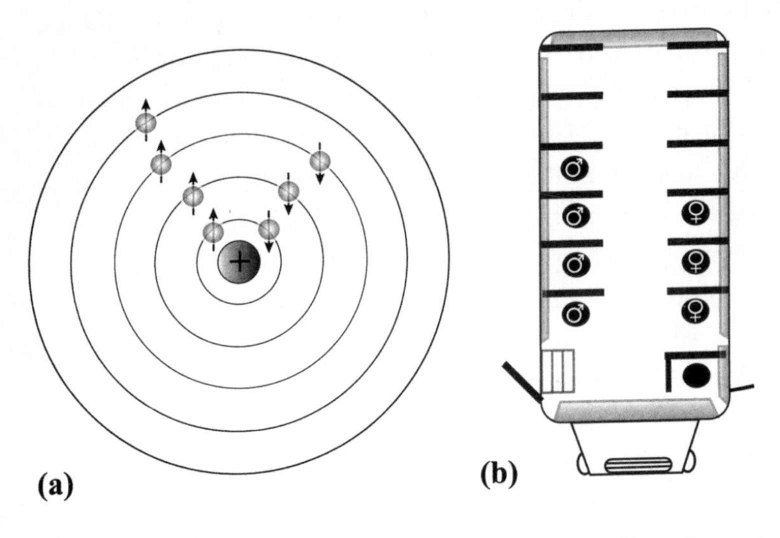

Figure 19.2 (a) Each allowed state or orbital of electrons revolving about an atomic nucleus can be occupied by only two electrons—one of spin up and the other of spin down. For heavier atoms, as more electrons are filled in, they are forced to occupy higher and higher orbitals. (b) The situation is very similar to people filling in the seats in a bus; people usually fill in the front seats first, and as more people come, they start filling in seats more and more toward the back. If the bus happens to have only one seat per row on either side of the aisle, and also happens to operate in some conservative society where men and women sit on opposite sides of the aisle, the similarity is even closer, if we compare men and women with spin up and spin down.

This exclusion principle is the basis of all of chemistry, which hinges upon how electrons inside an atom are arranged in different quantum-mechanically allowed orbits about the nucleus. As we discussed in Chapter 1, electron orbits can only have certain particular radii (or distance from the center of the nucleus).[30] Now if we start filling in these orbits with electrons, then each orbit can have only two electrons, one of spin up and one of spin down, because of the exclusion principle. So once we have two electrons in, then the next electron is forced to be in the next higher or larger orbit, and so it goes on as we fill in more and more electrons. This is shown schematically in Figure 19.2(a).

This is sort of like how bus seats fill up—people tend to occupy the seats at the front first, closest to the driver and the entrance. If the bus were to have only one seat per row on either side of the aisle, and also happened to operate in some conservative country where men and women sit on opposite sides of the aisle, then the similarity can be striking, as we see in Figure 19.2(b): Each row can be occupied by one man on the left and one woman on the right, just like a single spin up electron and a single spin down electron in each orbital!

So much about fermions, what about bosons? The gregariousness of bosons actually leads to some very interesting behavior with important real-life applications. Because photons are bosons, we can put as many of them in the same region of space as we want with all in the same state, and when we succeed in doing that very efficiently, the result is something quite amazing that we all know about: laser!

---

[30] The allowed states of the electrons (as discussed in Chapter 1) orbiting the atomic nucleus depend on other factors besides the size of their orbits, such as angular momentum that defines how fast and in what orientation each electron is orbiting. But the basic idea is the same, each allowed state called an "orbital" about the nucleus can be occupied by only two electrons—one of each spin. To keep it simple, here we will just think of each distinct state being defined only by the size of the radius of the orbit as shown in Figure 19.2(a).

Earlier in Chapter 1, we mentioned the wave-particle duality, a lingering mystery in quantum mechanics, which indicates that all particles in the universe, fermions or bosons, behave both as particles and as waves, depending on the situation or experiment involved. So it is with light—which is comprised of particles called photons and also can be understood as electromagnetic waves. We can visualize the situation as is shown in Figure 19.3. In ordinary light, the photons, although crowded together, are not quite in sync with each other, as shown in Figure 19.3(a). The waves associated with the photons are in diverse orientation, and there is a lot of destructive interference among the waves (see Chapter 9), and this results in relatively low intensity and very diffuse light. In laser light, on the other hand, all the photons are in identical states, with their associated waves oriented the same way and in generally constructive interference, as shown in Figure 19.3(b). This results in very efficient and high-intensity, focused light. The difference is easy to see: Ordinary light can light up a whole room, while laser light creates a single very bright spot, because all the light is oriented the same way and therefore focused toward a very specific direction in space. In lasers, the bosonic nature of light truly becomes manifest, since it results from zillions of photons essentially being in the same state. If photons were fermions, lasers would have been impossible.

Now, one might say dismissively, "Come on now, this is not so surprising, after all photons have no mass, they have no material substance, so of course we can pile them into the same state." True, but then I need to tell you about the Bose-Einstein condensates (BEC). Something very similar to what happens in lasers can be made to happen with atoms as well, which are the building blocks of all matter, including human beings. But that can happen only at

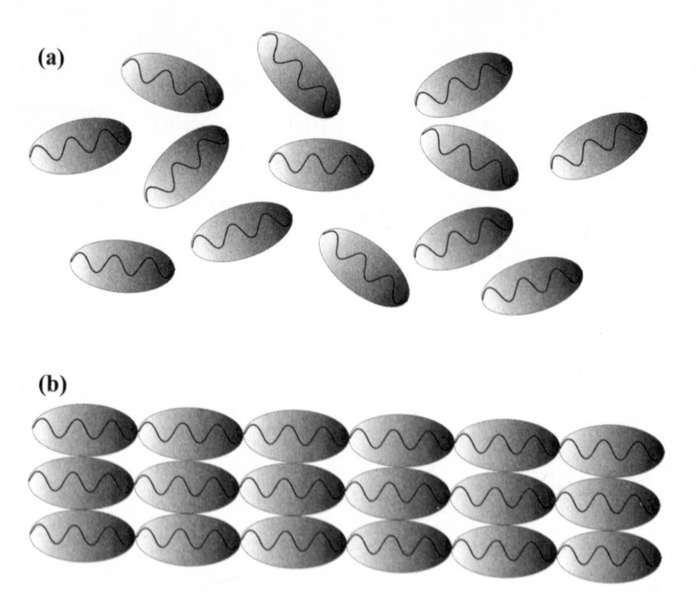

Figure 19.3 Wave-particle duality of light is represented by little ovals for particles containing bits of waves in each of them. (a) In ordinary light, the waves may be oriented differently and may be out of phase, so that there is a lot of destructive interference among them, as discussed in Chapter 9, resulting in reduced intensity and rather inefficient light. (b) In laser light, all the waves are oriented the same way, and they are almost all in phase with each other (meaning crests and troughs of the waves are aligned), and this results in very high-intensity light.

very low temperatures, close to absolute zero;[31] creating BEC in the laboratory was one of the major triumphs of experimental physics in the last decade of the twentieth century. The idea of BEC has a long history: It was first predicted by Einstein in the mid-1920s, but

[31] Absolute zero is the lowest temperature possible. It is approximately–273.15° Celsius or–459.67° Fahrenheit. As mentioned in Chapter 3, temperature relates to the disorderly or random motion of atoms and molecules. At absolute zero, all such motion stops (from the classical physics perspective), and the atoms come to a standstill; therefore, no further lowering of temperature is possible. However, absolute standstill would imply precise knowledge of the atom velocities (exactly zero velocity), which conflicts with the uncertainty principle discussed in Chapter 2; therefore, in the quantum view, there is always some residual movement even at absolute zero, appropriately called *zero-point motion*.

it took seventy years until it was finally created in the lab in 1995, because of the challenges of creating such low temperatures. As the BEC experimentalists like to brag, the coldest spots in the universe are right here on earth, in their labs!

But how does a BEC happen? At room temperatures, atoms behave as in Figure 19.4(a)—like well-distinguishable particles with tiny wavelength (see Chapter 9 for its definition). As temperatures get lowered, the wavelengths of the atoms get longer and longer[32] until waves sort of merge with each other, as shown in Figure 19.4(b), and all the atoms behave like one single entity! That's a BEC or Bose-Einstein condensate.

While there are obvious similarities between BECs and lasers, it is worthwhile to point out some key differences. First, lasers obviously work at room temperature—we use them all the time in optical drives, DVD players, and laser pointers—but BEC requires ultracold temperatures near absolute zero. That's partly because the wave-like properties of atoms become manifest only at those very low temperatures. Second, a BEC usually stays in one place—where it is created—unless it is physically moved by some complicated machinery, but laser light, like all light, can never truly stand still and is always moving at its very high velocity.

Lasers and BEC simply reinforce certain essential qualities necessary for human success but so hard to harness: discipline and teamwork. One can think of lasers as an incredibly disciplined bunch of photons moving together as a team in a precisely coordinated way—sort of like a highly disciplined army. The military and successful businesses are known for being disciplined and for coordinated teamwork, but both discipline and teamwork do not

[32] By the way, such quantum mechanical waves associated with material particles like atoms are called *matter waves* and their wavelengths called *de Broglie* wavelength after an aristocratic French physicist who first suggested the idea of matter waves in 1924 in his doctoral thesis, for which he got the Nobel Prize.

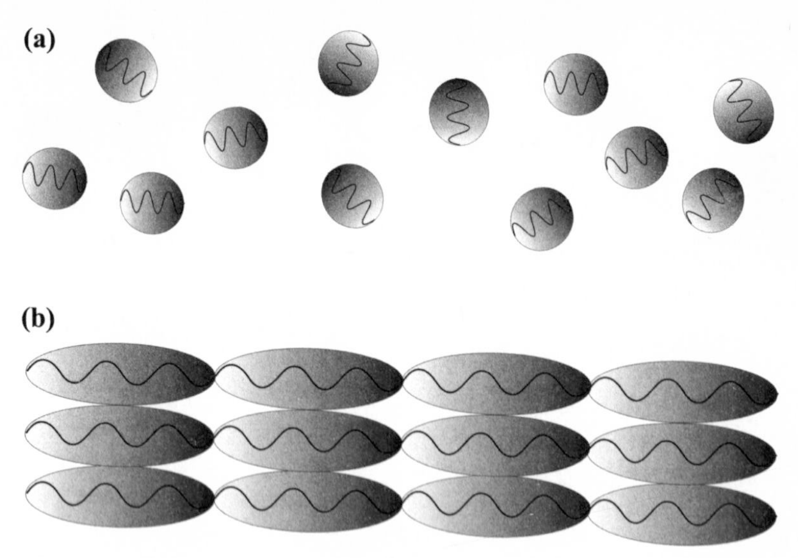

Figure 19.4 (a) In an ordinary gas, atoms (both fermions and bosons) are all in different states, moving around at random. At room temperature, for example, their wave-like characteristic is strongly suppressed, meaning the wavelengths are really small, and the particle nature of the atoms is dominant. (b) At really low temperatures (close to absolute zero), the wave nature of the atoms becomes very prominent, and wavelengths get longer, and if the atoms are bosonic, then they would all settle into the lowest energy state. When that happens, all the atoms sort of blend into each other as shown, and they all behave like one gigantic mega-atom comprising sometimes millions of individual atoms. This is an exotic state of matter, called a Bose-Einstein condensate (BEC). Notably, fermionic atoms cannot achieve this, because they cannot all be in the same state.

usually come naturally to us and require particular effort. Well, so does creating laser light even though it involves bosons. Perhaps we can blame our difficulties with teamwork and discipline on being constitutionally made of fermions that prefer to be by themselves! But as far as our individual personalities go, there is indeed a broad range, and so you might ask yourself, are you bosonic or fermionic in nature?

CHAPTER 20
THE MIDDLE PATH INTEGRAL OF LIFE

The doctrine of the Middle Path is central to Buddhism and provides one of the most valuable and enduring bits of wisdom: that moderation is the key to stability and happiness. There is a mathematical reason for this that can be found in the idea of path integrals and the related quantum notion of sum over histories.

In any romantic relationship, there are periods of delirious happiness, when you feel at the top of the world, and everything is picture perfect. That certainly has a tendency to happen more often near the beginning of a relationship, when your object of affection seems to be the embodiment of perfection, and you love just about everything about him or her. But after the honeymoon period is over, as the patina of perfection wears thin with growing familiarity, and those cute eccentricities start to become irritating behaviorisms, then inevitably there come periods of frustration and despair. Love, romantic love in particular, is just like that; it can make you so happy to be alive, making life worthwhile, but it always carries with it the unwanted but recurrent bouts of despair and depression, when none of it seems to be worth it. And we all know quite well that this transcends the complications associated with love and relationships. It actually illustrates a fundamental fact of life: that anything that can bring us great joy also has the potential to bring us great sorrow; anything that gives us intense pleasure can also cause us equally intense pain. As an example of the latter, just think about alcohol or any of the illegal recreational drugs.

That's life: All the important and essential things, all those things that bring fun and joy into life, all fit the description of "can't live with it, can't live without it." It's a tough conundrum. But a while back I realized something interesting about this ubiquitous catch twenty-two of life, something that has made it all quite bearable and actually downright comforting at times. Let me share it here with the hope of encouraging a more positive attitude while facing the rough times, as we all inevitably have to sooner or later.

The general idea can be best visualized with the concept of *path integrals*, an essential tool in physics and math. If you ever took calculus in college, you might already have seen more integrals than you bargained for. But if you have been getting along in life just fine without any calculus, that's no problem, because you really do not need to know any calculus to understand what a path integral is. It simply involves adding up all the values that a parameter or variable of interest takes along a certain path. Let me illustrate. Say you went biking one fine Saturday afternoon and did a 5K circuit that took you over a somewhat steep hill. Your bike has a GPS device that tracked your elevation at intervals of 100 meters along your route. At the end of your ride, the GPS gives you a chart that might look like the one in Figure 20.1, where the little circles mark out the measured elevations on your route. Now if you were to add up all of those elevations from the beginning to the end and then just divide the total by the number of measurements or points on the chart (fifty in this case), you would have just computed an approximate value of the *path integral* of the "parameter of elevation" along this particular path. As you might have guessed already, the path integral in this case is just a measure of the average elevation along your bike route; and *that pretty much is what simple path integrals do—they compute the average of some parameter along a specific path.* However, we have to keep in mind that this average (equal to 27 meters in Figure 20.1) is not an exact value for the path integral, but only an approximation. But we can always get closer to the exact value by tracking the elevations more frequently. Say instead of noting the elevation every 100 meters, the GPS is set to mark it at every meter; we would then have 5,000 points (instead of 50), one for each meter of path, and they will be closer together. The resulting average will give a much better approximation for the path integral—but it is still not quite exact. So how do we get the exact value? This is where calculus comes in. Let's imagine taking this to the limit of a continuum—suppose we have a magic GPS that measures the elevation at every infinitesimal point on the path, continuously. In that limit, the average we get becomes the exact path integral, which corresponds to the exact average elevation during this bike trip. You might wonder, "Since there are an infinite number of points on that path, wouldn't the sum or integral be infinite as well?" But remember we are also dividing by the total number of intervals, and there are infinitely many of them as well, which compensates for it, so as a result even in the continuum limit the path integral remains a finite number not very different from the first approximate value of 27 meters we got from the chart in Figure 20.1.

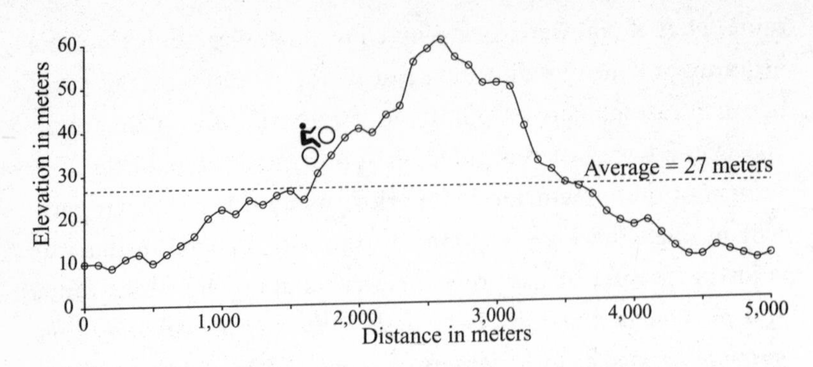

Figure 20.1 Example of a simple path integral: During a bike ride, if we add up the elevations measured at fixed intervals and divide the sum by the total number of intervals, we would get the average elevation. If the intervals are made smaller and smaller with more and more of them, the number for the average elevation gets more and more precise. In the limit of measuring at all points (and infinitely many of them) along the path, the average is exact, and we have a path integral.

With the idea of a path integral on hand, we can now size up this life conundrum that we were talking about—that everything that brings us joy and pleasure can also bring us sorrow and pain. Experience tells us that there is no getting away from that. But there is a bright side to this that we almost never appreciate or even realize. Despite all those ups and downs and pros and cons of everything that we call life, it turns out that for most people, the path integral of joy/misery over their lifetimes remains within a very narrow and comparable range. What does that mean? Here's how it works. Say we assign a positive number to the joy or happiness a person feels at any time, on a scale of 0 to 10, and a negative number to the level of sadness on a scale of 0 to –10 (after all, sadness is the negative of happiness). This is just like the scale with the cartoon faces we see at a doctor's office to help us communicate the level of pain we might be experiencing; the only difference in our scale, shown in Figure 20.2, is that we are now rating the level of happiness as well as sadness. On this scale, 0 means neither sad nor happy, 10 would be the happiest one could ever feel, and –10 is the absolute nadir of sadness and depression. Now suppose we have a *mood-meter* (like the GPS in the bike ride) to track a person's mood at regular intervals and mark them up on a chart. In Figure 20.2, we see an example of such tracking done daily for someone for fifty consecutive days. If this particular person seems to be a moody sort with strong fluctuations of mental state daily, that's just to make the point and is not necessarily the typical scenario. The chart is pretty self-explanatory: Each dot on the chart corresponds to a value between +10 and –10 as measured along the vertical axis on the left. The thin line connecting those dots makes it easier to track the changes. As you can see, this is very much like the elevation chart in Figure 20.1 that we drew earlier, and just as we did there, if we average over all the points along this path, we find an approximation for the path integral of happiness/sadness over those fifty days for this particular person. As with the example of the bike ride, we could in principle take the continuum limit by tracking the person's mood at every instant and get an exact average or path integral of happiness/sadness over that period of time. But notice, and this is important, although the person's mood and state of mind fluctuates between all the way up at +8 and down close to –8, the average (or path integral) of happiness/sadness is just about 1.5, as indicated by the dotted line in Figure 20.2. Let us call such path integrals over any period of time the "net-joy" value for that person for that particular period.

It is clear that we can define a net-joy value for our entire lives, as well. Now we might tend to think that if we chose certain paths in life, this net-joy would be close to a stellar 10, implying a uniformly happy life, as opposed to some other path in life that we luckily avoided that would have led to a net-joy value depressingly

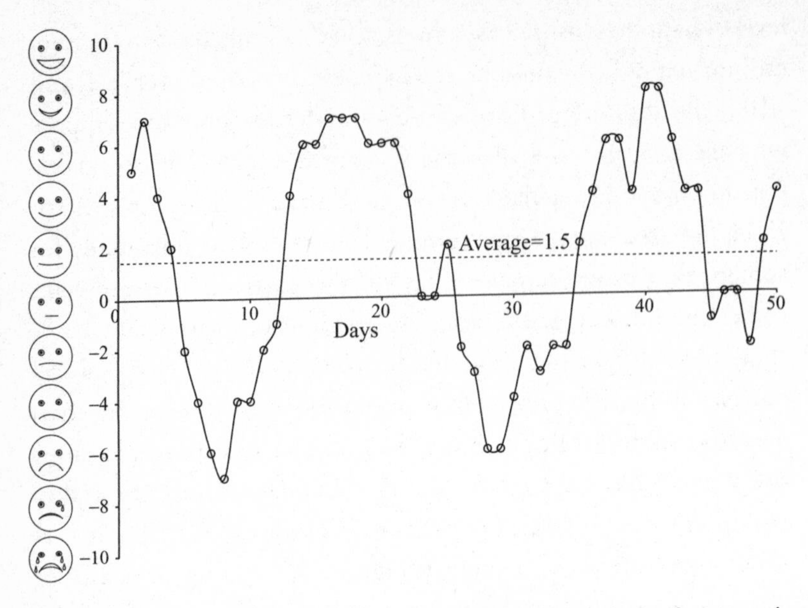

Figure 20.2 Path integral of happiness/sadness: Keeping track of a person's level of happiness/sadness daily gives his or her average mood, shown here for a period of fifty days. If we can track the person's mood at every instant, that average will be the path integral of happiness/sadness of the person over that period. Over a lifetime, for most people the average remains close to zero, as illustrated here even for a short period.

close to −10, implying a persistently unhappy and miserable life beset by rotten luck. But in reality, I claim that, for *most* people, the net-joy value will be a number rather close to 0, regardless of what path they choose to take in life. Skeptical? Let me elaborate further.

Say you decided to play it safe, took no risks, emotionally or otherwise; the likelihood is that you will never feel deliriously happy, but then you will also probably have few periods of extreme grief and despair. So on the happiness/sadness meter, you will oscillate perhaps between about +2 and −2. But instead, if you went all out, and decided early on that you wanted to experience life to the extreme, and you risked your heart repeatedly in hopeless romantic propositions, took chances with your careers, indulged in extreme sports, took high-stakes gambles, then you will probably have moments of sheer exhilaration, you will hit +10 often. But it is also certain that you will equally often hit –10 as well when your heart, mind, and body are battered by those risky ventures you undertook. So, when you add it all up, the sum is perhaps the same as what you would have had if you played it safe. But of course on this second route, you can claim to have experienced a lot more of life—drunk of life to its dregs, as they say—and that is often perceived as a positive thing. Well, at least it makes for a better story to tell. If, as The Biography Channel claims, every life has a story, the second route would make a bestseller.

However, it should be quite clear that the path itself may vary; there are many possible paths everyone's life could potentially take, based on the interplay of luck and choices—the infinity of world lines that we talked about in Chapter 17. What remains more or less constant is the net sum or the path integral of happiness + sadness over the entirety of each of those possible paths. In the end, we all end up charting one unique path as we go through life and make our choices, the path that we call our life story or biography. But for most people the ups and downs in life mostly average out regardless of what that path is. Perhaps this simple observation can help us relax somewhat, as we all inevitably ponder the eternal "what if"s of life and are sometimes tormented by visions of the paths not taken. Because no matter what path we might have taken, the sum of happiness and sadness would probably average out to be close to what we have right now. If we chart our lives to experience extremes of sensation, it is almost inevitably negated by the opposite extremes. Things that make us deliriously happy will inevitably make us feel like shooting ourselves at other times.

Everything we enjoy in life follows this rule in one way or the other. Relationships are one obvious example. It is much the same with kids; they bring a lot of joy into life, but as any veteran parent will testify, they can also drive you nuts. Or take food for that matter. If you are into the delights of spicy food, the farther you climb up the Scoville scale of spicy heat, the more likely it is that you will be sweating it out on the ceramic throne the next morning. The illegal recreational drugs are perhaps among the more extreme examples. They apparently take you to whole different levels of pleasure and ecstasy. But we all know from the horror stories we hear of and see, in life and in the media, how drug addictions typically end—the pain and misery match the thrills felt at the beginning.

But of course, in life there are always exceptions to every rule, and the zero-sum scenario is not always true. Seriously extreme behavior can break off the ups and downs early on in life, and it might end up prematurely on a high or a low. When some go to (or life takes them to) serious extremes, they might just go off the deep end of depression with a gun to the head and never come back, or even get off on the highs of ecstasy and die of an overdose of something. At the other extreme, a lucky few might be elevated to a string of just good luck and fortune all through life. But generally speaking, for most people, the good and the bad do manage to average out.

Taking a philosophical perspective of an entire lifetime, while in the middle of one, is not an easy task, since we mostly live in the here and now. So, if you want a validation of this rule of life playing out in real time, just consider those moronic drivers (well, perhaps we all have been one at times) who weave in and out of heavy traffic endangering their own lives and others'. Most of them believe that they are getting to their destinations faster than the rest of us. Well, they usually don't, and it often happens that you run into those guys again later on at some traffic light, because it has taken you exactly the same amount of time to get there. They might go faster over stretches and then inevitably get bogged down more at other places. Well, if they go too fast, you might even pass them by as they pay the price for the thrill of their flash of speed, parked in front of flashing lights. That's pretty much what happens in life: The paths that are steady might not have the exhilarations of high speeds, but they avoid the dangers and pitfalls that come with the exhilaration.

What I am saying here is nothing new really, just put in a different light. Someone infinitely wise anticipated me with a similar conclusion two and a half millennia ago. Gautama Buddha realized this fundamental truth about the conservation of happiness and sadness in life as a central tenet of the general enlightenment he experienced. That is why he recommended the Middle Path. The Middle Path is all about reducing the happiness/sadness oscillations in life. The Middle Path is all about avoiding the extremes of life. That way, we avoid the abysses and the pitfalls, but the tradeoff is that we miss out on views from the peaks and mountaintops of life, as well.

Buddha's quest for the truth about life started with his encounters with human misery. A protective father, so the story goes, tried to shield him from the harsh realities of life. Ironically, this only accentuated the impact when he inevitably saw death, the infirmities of old age, disease, poverty, and everything else that is the lot of humans. Prince Siddhartha, as he was known then, renounced his kingdom, his wife, and his child to search for a way to minimize the impact of these sorrows in life. If we look beyond the mysticisms of afterlife and the recurrent cycles of life, the practical solution that Buddha came up with, to a large part, amounts to recognizing this essential conservation of the happiness/sadness sum in life and that only by avoiding the extremes can one with some certainty avoid the high-intensity despair and disasters in life.

Moderation is not just a mystical Buddhist principle, it is actually the only proven and time-tested way to optimizing the path of life. That includes health, happiness, and peace of mind. We are always dazzled by the extremes of joy and glory, pleasure and happiness. We are good at suppressing the flip side, the extremes of depression and sickness and sadness that exist out there as well. Just check out some of those archival *Behind the Music* shows on the VH1 television channel to get a reality check if you ever suffer from "rock-star envy"—it certainly cured mine! The fall is quite often just as hard as the heights climbed.

The quantum worldview has something even more intriguing, albeit even more comforting, to offer. If we were to interpret Richard Feynman's path integral formulation of quantum mechanics quite literally, we could come to the conclusion that there is really no point at all in worrying about what path we take in life—because, as a matter of fact, we are actually taking ALL of them! Well, if you are muttering, "Say what?!!" let me first mention that this formulation in terms of path integrals (in a somewhat broader sense than we have seen here) is perfectly legitimate and is mathematically equivalent to the standard formulations of quantum mechanics; therefore, it is absolutely valid in the realm of subatomic particles. In fact, it is indeed the most evocative and intuitive way of understanding quantum mechanics—and makes one wonder a bit whether our lives work out the same way as well. Although a bit hard to accept, it is easy enough to understand the main idea, using the perspective of life outlined in this chapter and also aided a bit by the vision of *world lines* of life from Chapter 17, which can now be interpreted as just different possible paths in our lives. Everyone's life has two certain events: birth and death. But there are infinite

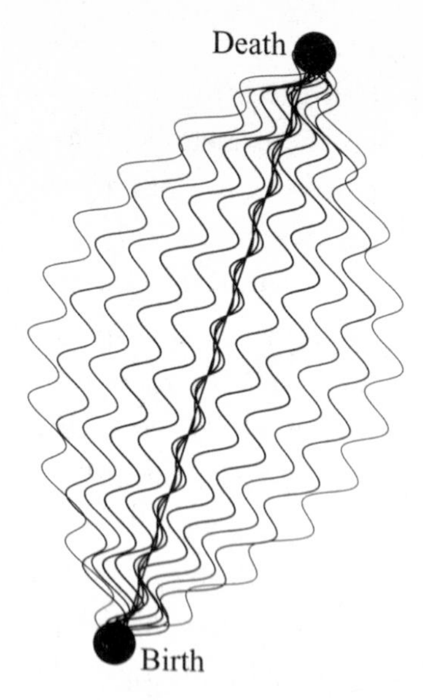

Figure 20.3 In a path integral vision of life, we might actually simultaneously be taking all possible paths in life. It is just that the probabilities of most of those paths cancel out like waves in destructive interference. Only for some paths (shown in dark in the middle) the net probability is significant, and they are the ones close to the reality of which we are aware.

possible paths that connect those two fixed points of destiny of each life.[33] Feynman's path integral point of view is that everything in the universe, including you and me, can be viewed as traversing all of those infinitely many possible paths or "histories" from one fixed point (birth) in time and space to the other (death)—sort of like being simultaneously on all possible world lines of life. Of course, the obvious question here is, if we really are following infinite possible paths of destiny in life, how is it that we are aware of only one? Well, as mentioned a few times before, in quantum mechanics, everything in the universe is described by waves of probability. But waves interfere with each other constructively or destructively, as described in Chapter 9; so that if two waves have their crests lined up, then they add up to interfere constructively to result in waves with a bigger crest, but if the crest of one lines up with the trough of another, they interfere destructively to effectively cancel each other out. So essentially what happens

[33] To keep it simple and closer to the ideas in physics, we will assume that the time and place of the starting and ending points are truly fixed and unchanging. If we allow for flexibility in the time and place of birth and death, that just opens up more possible paths, which does not change our general conclusions here in any way.

in this path integral worldview of quantum mechanics is that we add up the waves of probability of all the infinite possible paths one could take. When we do that, the probability waves for practically all the paths (except in the vicinity of the one we actually follow in life) undergo destructive interference with one another, canceling each other out. Only close to the actual path we follow in life do the probability amplitudes add up constructively, resulting in the true reality in which we live out our lives. The general idea is sketched in Figure 20.3. The vision this evokes has a hard-to-accept dizzying madness about it and yet has a poetic beauty to it at the same time.

While expressed in terms of human lifelines, it may seem like science fiction, conjuring up alternate histories and parallel lives. It is possibly just metaphorical (but who knows?). But, in the realm of subatomic particles, including those that we are made of, this describes reality very accurately, and it is an absolutely valid viewpoint that serves as the basis for perhaps the best understanding of quantum mechanics that we currently have. If we extend and accept that viewpoint to be true in life, as we accept so many other things, perhaps we can stop wondering "What if?" and make the most of the one path of which we are aware.

CHAPTER 21
A CURRENT FORMULA FOR SUCCESS

We finally have a real mathematical formula for success that takes its cue from natural laws. Success is about reaching our goals and getting to life situations we want to be in, which means that it is all about harnessing the currents of life and channeling them in the right direction. There is a simple physics formula that describes the flow of currents, and its human interpretation contains the essential ingredients for success.

These days we hear a lot about "formulae for success." Motivational speakers and authors of self-help books rake in millions giving advice on how to reinvent and reorient yourself for success. For a lot of people in their target audience, success is mostly about making more money, but I have a strong suspicion that the only people who ever get rich from those ideas are the ones selling them. That aside, success can mean a lot of different things for different people, but whatever that might be, we all recognize a success story when we hear one because they all have something in common: Success is about attaining goals and achieving life situations that we wish for. It might be more money, more free time, better quality of life, stronger marriage, happier relationships, fame and recognition, satisfaction in a job well done, the list goes on. And if we happen to beat incredible odds getting there, it is so much the better; it might even lead to a movie deal!

As we start out in life, most of us take it for granted that success will come our way. The general attitude in youth is that "other people might fail, but not me." Success? Absolutely! There is plenty of time for it all to happen in the future. Well, time marches on, life happens, and then for most people, such optimism eventually starts to fade as the realities of life set in, until one day we realize that the future is now. Then a cold suspicion starts to creep through our denials that perhaps all those dreams we had will remain just that—dreams. Most people adjust to the new reality by severely lowering expectations and becoming "comfortably numb," to quote Pink Floyd. But always there are some stubborn souls out there who refuse to let go of their dreams, and a desperate search begins for quick fixes—for some formulae to set them back again on the path to success. They just became a part of the target audience for all those motivational speakers and the self-help books.

It is particularly painful in these connected voyeuristic times when we are constantly bombarded by the media with the minutest details of the fabulous lives of celebrities, the rich and the famous, all those apparently successful people. We are mesmerized and just can't help watching, and we get clear and vivid visions, in color and in motion, of how we would have liked our lives to be, and we ruefully look at the reality of our own lives—so far from all that. As if that is not bad enough, social media sites like Facebook daily rubs into our faces all the smiling happy pictures of those hundreds of "friends" chilling out at parties and in exotic locations, which so successfully mask the routine reality of their lives that we are left thinking that everyone else is having a great time but our own miserable selves. As a result, nowadays we are rapidly morphing into a culture that is just frustrated and dissatisfied often for no particular reason.

But that's the world we live in. So unless we can shut out all media and achieve supreme levels of detachment *à la Buddha*, we will always be looking for something more, to be somewhere else, and we could all use a formula for success—success in getting what we want and where we want to be. Even the most successful people cannot be successful on all fronts in life. Therefore, regardless of who we are and where we are in life, we could all use a surefire formula for success.

That phrase, "formula for success," is thrown around often and quite cavalierly. But the word "formula" evokes equations and mathematical symbols, yet there is nothing really scientific or mathematical about the advice you get from the self-help gurus who relieve you of your hard-earned money. But what if we could indeed come up with a mathematical and scientific formula for success? Considering that we want a formula for success to improve the chances of getting to our goals, the hard sciences have just the formula we are seeking.

A scientific solution requires a precise definition of what success means to start with. I think we can all agree that success is about reaching our dreams and goals. Therefore, success is about making progress in life in the direction we want to go, which means success is really all about *flow*—strong sustained flow in the right direction. Too often in life, we lose that flow, and we literally stall and stagnate or even go into reversal. We lack or lose the motivation to do what we need to do to get to where we want to be. To snap out of it, to get ahead, to get to those dreams, we need to jumpstart the flow and channel it along the right path; we need to put ourselves in charge of that flow. Now, here's the good news: Flow is something physics knows a lot about; physicists have been studying flow for the longest time, and there are precise equations that define flow. The human interpretations of these equations contain the prescription for success.

The defining characteristic of all kinds of flow is the *current.* A stronger, larger current means a bigger flow—think of flowing water, for instance. In physics, understanding currents is among the most important tasks for real-world applications. Water flow generates hydroelectricity; currents of air turn wind turbines; flows of hot and cold gases, air, and steam are used for heating and cooling, and to run engines; and most importantly, flow of electric charge through wires and circuits (electric current) is responsible for most of the technology we enjoy in modern life. Modern technology is based upon harnessing different kinds of flows in nature. But the truly amazing thing is that whether it is electric charge or water or hot gases, all kinds of flow are essentially described by the same basic formalism in physics. The details might vary from system to system, depending on the complexities, but the essential formula is the same for all; that is the way we would like it to be if our goal is to extract from it a universal formula for success that applies to everyone regardless of individual aspirations.

Since almost all kinds of nonturbulent flows and currents share the same underlying features, we have the freedom to pick and choose. So, to be concrete and to have a specific picture in mind, let us consider the case of electric current. In simple terms, electric current can be thought of as the directed flow of tiny little *subatomic particles* called *electrons* that carry *electric charge*, moving through a wire.[34] Electric charge is the stuff electricity is made of; thus, when we "charge a battery," we build up more electric charge

[34] The quantum view is somewhat more sophisticated, but the end result is quite the same.

at its terminals. There is a simple formula that describes the flow of these tiny particles responsible for electric current, and here it is:

$$J = v \times n \times e$$

This simply means that the current (that we denote by J) can be obtained by multiplying three characteristics upon which it depends. Stated in words, the formula reads:

***Current** = (velocity of the particles) × (density of particles) × (charge on each particle)*

It is easy enough to understand why and how the current depends on these three characteristics:

v (velocity of the particles): For any kind of flow to happen, the particles need to move, meaning they need to have a non-zero velocity, and a faster velocity would obviously lead to a stronger flow.

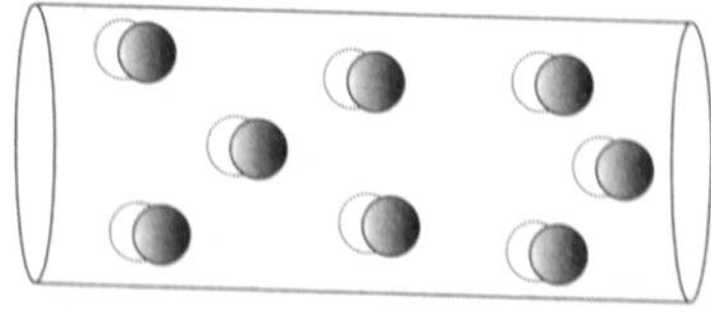

Slower particle velocity means weaker current

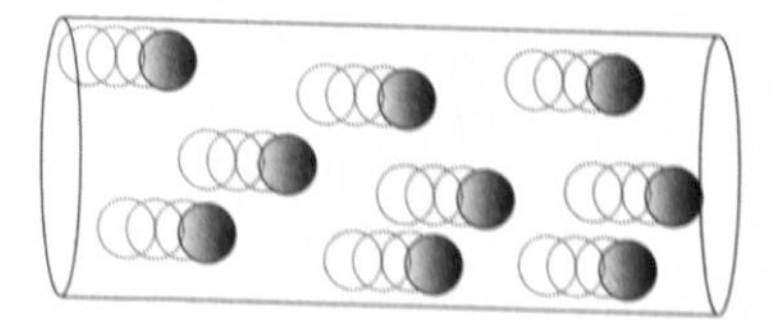

Faster particle velocity means stronger current

Figure 21.1 Higher density of particles generates a stronger current.

n (density of particles): The strength of the current would most certainly depend upon how many particles are available to flow in the first place. In the limit of no particles at all, clearly there would be no current! Therefore, it follows that a higher density of particles going through generates a stronger flow and vice versa.

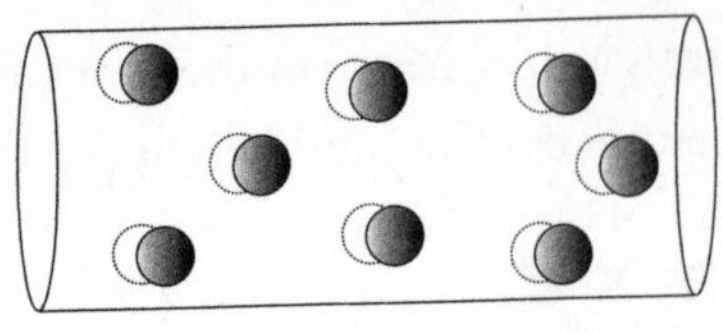

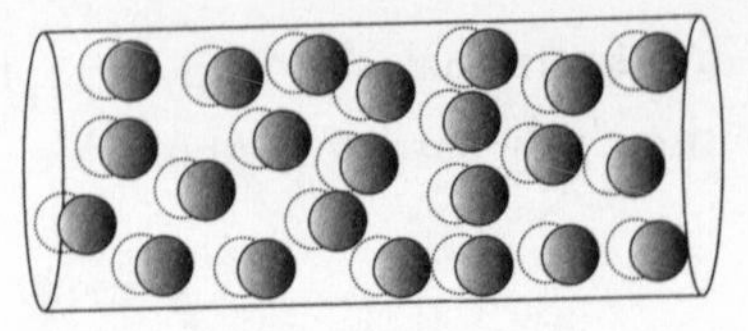

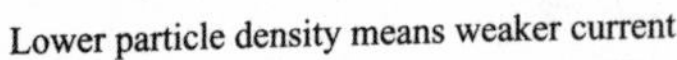

Lower particle density means weaker current

Higher particle density means stronger current

Figure 21.2 Higher density of particles generates a stronger current.

e (charge on each particle): The strength of electric current must certainly depend on how much electric charge (amount of electricity) is carried by each particle. If the particles carried no charge at all, there certainly would be no electricity flowing. Well, for an electron, the electric charge happens to be a fixed quantity, but in other kinds of flow, each particle could in principle carry a larger or smaller chunk of whatever is flowing, leading to a higher or lower current accordingly.

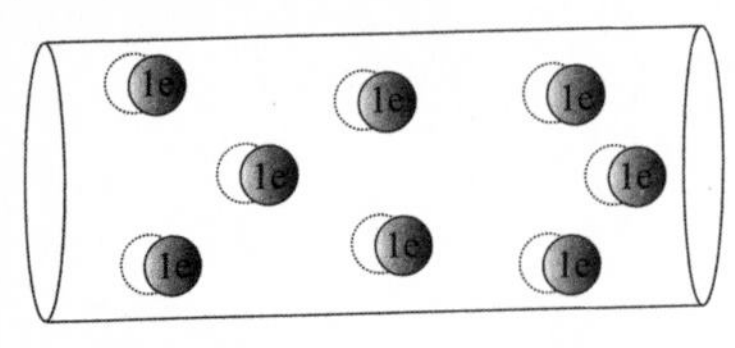

Smaller particle charge means weaker current

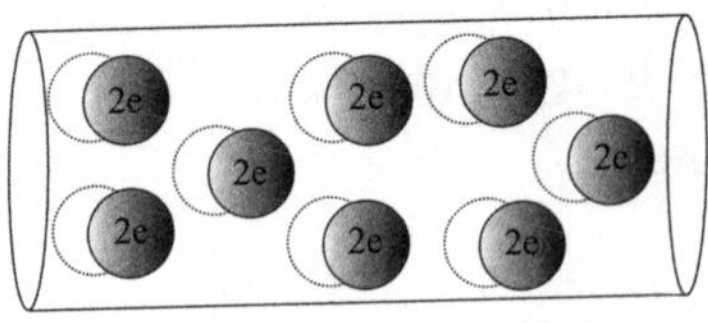

Higher particle charge means stronger current

Figure 21.3 If each particle carries a bigger electric charge (or a bigger amount of whatever it is carrying to generate the flow), the current is stronger. Here "1e" represents one unit of electric charge, and "2e," two units.

So it is clear that the current depends on these three characteristics, but why do we multiply them together? That is because if any one of them increases, the current increases in direct proportion. If, for instance, the charge density doubles, the current would be twice as large, or if the velocity is tripled, then the current would be three times stronger, and if both happen simultaneously, then the current increases by a factor of $2 \times 3 = 6$. Multiplying all three variables ensures that the current is proportional to all of them.

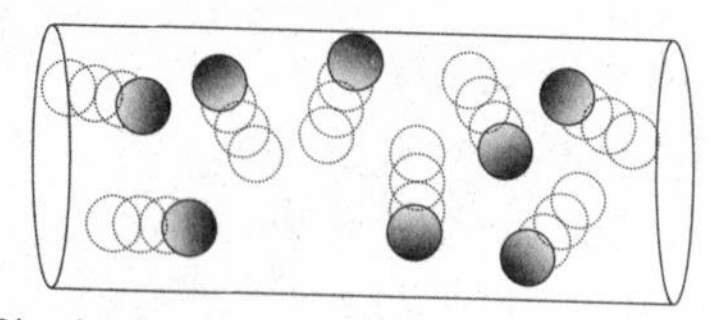

Disordered random motion means weak or no current

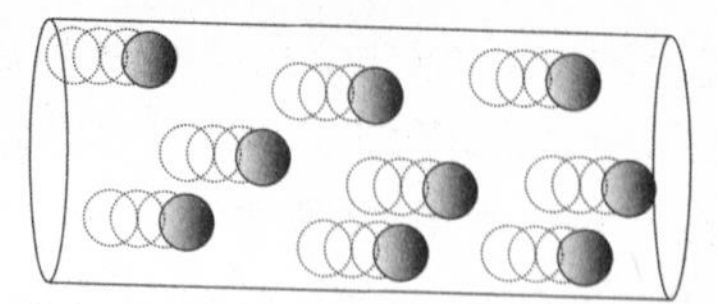

Ordered and directed motion means strong current

Figure 21.4 Ordered or disciplined movement is necessary for flow. The current is the strongest when all particles move only in the direction of the current. Completely random motion would give zero current.

But the formula as written above leaves out something very important in the physics of flow, something that plays a defining role in human success, as well. See, we tacitly assumed that the velocities of all the particles are in the same direction—in the direction that we would like the flow to happen. Of course that does not have to be the case—the particles could all be moving in different directions and even backward, as shown in the first picture in Figure 21.4.

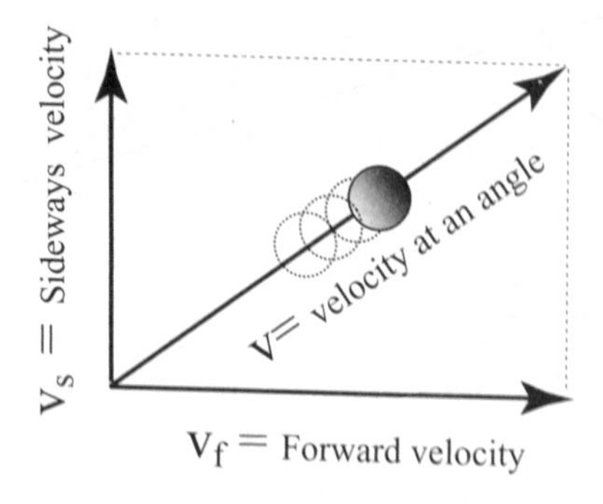

Figure 21.5 If a particle moves at an angle to the direction of flow, we can "break up" its velocity into "forward" and "sideways" velocities. The arrow-lengths are proportional to the velocities they represent. The forward velocity is always less than the total velocity because part of the total goes toward sideways motion.

But only the forward part of the motion can contribute to the current; sideways motion doesn't count, and backward motion would actually reduce the flow. So, for each particle, we should pick out and include only the component of its motion that is along the direction of the flow; if the motion is at an angle, we simply break down the velocity (as shown graphically in Figure 21.5) into a sideways velocity (v_s) and a forward velocity (v_f)

along the direction of the flow. Backward velocity will be ($-v_f$), the negative sign indicating that the particle is moving oppositely. But there is one more thing: Since the particles could all be moving in different directions, to obtain the net flow, we will need to take the *average* of the *forward component of the velocities* of *all* the particles. In physics, an average of a quantity is often indicated by placing a "bar" or horizontal line over its symbol, in this case, v. The upshot of all this is that we simply replace the velocity factor (v) in our original formula with the average of the velocity components in the direction of the flow, so our final formula for the current reads,

$$J = \bar{v}_f \times n \times e.$$

This might look almost the same as before, but the small changes we made make a very important point: *motion is not enough for flow; the motion has to be directed and coordinated.* Thus, if the particles were all moving in random directions, their movements might even average out to zero leading to zero current, just like people milling around a crowded room. On the other hand, if all the particles were moving in the same direction, there would be a significant flow in that direction, like if some dimwitted practical joker were to release a smoke bomb in a crowded room with a single exit and shout "Fire!"

That's all we need to know about currents and the physics of flow, but what about the formula for success that I promised? Well, there it is—the formula for the current is the formula for success; we just need to interpret it. Every element that appears in the equation for the current relates to an essential human quality absolutely necessary for success, and as we will see, it all makes perfect sense.

Velocity of the particles = Hard work: This is quite an easy and watertight connection to make because to get electrons, or anything else for that matter, to move requires work. If the electrons flowed together by themselves, then generating electric current would be free, and no one would be paying electricity bills. We all have to pay because somewhere work is being done (in a power plant) to get the charge-carrying electrons to move, and that movement propagates all the way to our homes via transmission lines as electric current to power all the appliances, lighting, and air-conditioning around the house. A higher velocity would increase the current, but that requires harder work. Well, as is often the case, conventional wisdom is distilled from truths of nature. As Thomas Edison famously said, "Genius is 1% inspiration, and 99% perspiration"; success boils down to sweating it out at the end. Without hard work, success is unlikely, just like the impossibility of having a current with zero velocity. Hard work is often much more important than just being smart and talented.

Coordination of the directions of the velocities = Discipline: We saw already that there is sustained current only if all the particles generally move in the same direction, the direction in which we would like the flow to happen. If they move around at random, then there might be no current at all. So the motion of the particles needs to be ordered and disciplined for better flow. Well, that just reaffirms the importance of discipline and focus to achieve success in life. We could work very hard and still get nowhere unless our efforts are disciplined and directed toward our goals, in quite the same way that particles could be moving very fast and yet lead to no net flow because they are all moving in different directions.

Density of carriers = Persistence: Just as a current requires lots of particles flowing together, in real life the flow toward success comes from a conglomeration of sustained bits of effort over time. The incremental motion of each particle is representative of each little step we take in our lives. If we never take any steps, we will never get anywhere, just like if there were zero density of particles, there would be no flow. Success usually happens only after persistent efforts.

It should come as no surprise that persistence is a part of the formula for success. Persistence should be a part of any formula for success. In fact, it is usually the most important ingredient. All too often, people fail because they give up too easily, where just one more try could have made all the difference. There is much truth in that old maxim, "Try, try again." The longer and harder we try, the better our chances are of eventually succeeding. If we scratch under the surface, we often find that most successful people had failed multiple times before they finally succeeded.

Persistence is important for success, but why do we relate it to the number density of carriers? Simply stated, the importance of a high number density of carriers for a strong flow is analogous to the number density of attempts that we make to achieve that success we crave. How many carriers or electrons there are available is really like how many times we try at something. A single electron flowing through a wire once every hour would not even register as a current in typical meters around the lab. To drive through a real measurable current, we need to have a certain critical density of electrons. In creature terms, think of one ant versus millions; one ant would hardly pose a threat to anyone, but in the tropical forests, people have been known to flee in terror faced with millions. We might think that this is more like one person versus several people. But we can turn this sideways—a task that could be done by several people simultaneously doing different parts of it could often be done by a single person doing it alone in multiple steps, one step at a time. Therein lies the analogy: If we tried just once and failed, perhaps we should have stayed at it, day after day, and eventually we would get it done, we would succeed. Persistence almost always pays off.

In case you did not notice already, it is significant that the current is not just dependent upon the total number of carriers, but on the number *density*. This is quite important, because I could have a million electrons in a very low density, spread out over a large distance, so that only one electron flows by every hour; that would be a quite negligible current. On the other hand, I could have just 100 electrons all clumped together in a high density so they all flow by in quick succession, and that would be a much more significant current during the duration of the flow. So it is the density that defines the current, not the total number. And so it is with our prospects for success; we need to keep trying again and again relentlessly and continuously. We cannot try once now and then try again ten years later; chances are that the opportunities that exist now will be long gone. It is our number density (or frequency) of attempts in a period of time that counts, not just the total number of attempts.

Charge on each carrier = Aggressiveness: This might seem a bit odd. Actually, to some it might sound downright negative. Nevertheless, it is an essential ingredient for success in most human endeavors. I am sure we all know people who work very hard and keep trying, have the right talents, and yet they always come up short with a bitter taste in their mouths, as they see less worthy and less talented people get ahead only because they were more aggressive.

Why should it be linked to the electron charge? Well, the charge of any particle is an intrinsic property, just like a person's aggressiveness. In electrical circuits, that charge is usually fixed; it is just the charge of the electron. However, in other kinds of flow, if we have a bunch of larger and denser objects (bigger "charge" in a more generalized sense), then of course the flow will be decidedly boosted. Same with aggressiveness—if properly used, it can clear obstacles in the path toward success. There is much truth in the saying, "Nice guys finish last," unfortunately quite true in all arenas of life really. But of course we never want to be *too* aggressive—we might turn people against us. However, it is amazing how a little bit of aggressiveness can get things done much faster than being timid and diffident, particularly when dealing with difficult people!

You'll note I did not list innate talent or skill here, or anywhere, in the formula for success—and with good reason. If you look around, you will find that for most people, their talent or skill had very little to do with their success. For some, yes, but the majority of people succeed with persistence, discipline, hard work, and with an aggressive attitude toward life and its challenges. Well, if we really want to, we could potentially take skills and talents and other intrinsic qualities and lump it together along with aggressiveness. But that is not so relevant in a formula for success, because most intrinsic skills and talents we are either born with or we are not, and we do not have much choice or control over that—all we can do is to develop them, and their development relies upon the four characteristics already in our formula. Without getting into the debate about "nature versus nurture," the simple truth is, if you do not have the bone structure of a super model, or the gigantic physique of a basketball player, or the superlative brain of Einstein, there is not much you can do about that. You can only develop the natural talents you have.

Well there you have it, a true formula for success:

Success =
(Disciplined Hard Work) × (Persistence) × (Aggressive Pursuit of Goals)

Really now, there is nothing here that you did not know already, but the beauty of it lies in its simplicity, because it takes nature's lessons to distill all of our conventional wisdom into four qualities that truly determine success. If this formula seems too simple, that is actually a good thing. Quite often people needlessly make things more complicated than they need to be so that they can justify charging money by filling up pages in books and hours in videos. You know, for many (not all) of the self-help and inspirational books out there, the message of the entire book can be gotten by simply reading the chapter titles, and even those are sometimes redundant! As a specific example, take the whole industry of losing weight; there are all kinds of weird diets and gizmos marketed, and people spend billions of dollars and hours, and really all it takes to lose weight for most people (unless there is a medical condition) is to exercise, to eat healthily, and to eat less. The trouble is most of us are inherently lazy and are usually hard pressed for time to exercise regularly, and we absolutely love the taste of unhealthy food (it is a rare food that is healthy and also tastes good!). So inevitably, we gravitate toward promises of miracle weight losses where we do not have to exert ourselves, and we are allowed to eat whatever we like and whenever we like—well, you can go ahead and spend your money on those gimmicks, but good luck getting any results!

But there are no gimmicks or frills in our formula for success—just the essential elements taken straight from the source of all our wisdom: nature. This formula for success is the real deal; it is a real mathematical formula, and just like that bit about diets and weight loss I just mentioned, the key to its effectiveness lies in its simplicity and universality.

AFTERWORD

Now that we have touched on many of the fundamental laws of the universe, we might well ask, "What makes a law of nature 'fundamental'?" If a law applies everywhere in the universe, and is something that is part of the essential truth of how our universe works, then it is a fundamental law of nature. Such laws are a part of the very fabric of the universe and are eternal and unchanging. We have encountered some of the most important ones in this book.

Although science can seem so overwhelmingly vast and complicated, it is useful to realize that there are not that many really fundamental laws of the universe that we are aware of, and much of the complexity we observe lies in the details. It is kind of like all the millions of different buildings, bridges, and architectural structures out there, but they are all mainly built of bricks, cement, steel, and glass—the fundamental elements are few and simple.

It is much the same way with our lives and the human experience. Our individual stories, personalities, and paths through life can be infinitely diverse, but permeating all of them there are a set of simple rules and patterns that define all of our lives. That's why we can instantly relate to tales of people separated from us by seemingly vast chasms of time and space and circumstances. Great literature, art, and music tap into those common and universal themes.

Most of us are instinctively aware of the existence of fundamental laws—those that define the material universe that we physically inhabit and those that define our conscious existence. This book shows that the two sets of laws have much in common, contrary to what we generally believe. But really, that should not be so surprising, because despite our natural preoccupation only with events and people that directly affect us, we are very much a part of the physical universe.

In the pages of this book, we covered the whole spectrum of issues relevant in life—relationships, sex, shared quirks of human nature, our need for company and privacy, our desire for fame, global flow of people and culture, how we manage our finances, how we perceive others, the sources of our common frustrations, finding success and the reasons we fail, how our sense of optimism and sense of time can evolve, what defines happiness, and so on. In the same span, we have also covered a broad range of the most important established laws of the universe. At every step, we find parallels—even when the physical laws may not directly apply to life, there are always counterparts in life that follow the same pattern. That fact is indeed nontrivial and profound, because it certainly did not have to be that way; the rules of life could have been quite the opposite of the physical laws. But they are not. The rules of life generally follow their physical counterparts. The implication is much more than a validation of the premise of this book—it tells us that at a deeper level, underneath all the subjectivity and bewildering craziness of life, we can always find bedrock of objective and concrete rules to guide us, if we only look for them in the right place.

Viewing what we typically consider quintessentially human behavior and concerns from the perspective of fundamental laws of nature therefore has the unique advantage of providing a certain objectivity that is hard to find amidst the unpredictable and often irrational course of life. Since this objectivity derives directly from the unchanging laws of physics, we can feel secure in the conclusions we draw from it and the worldview we build upon it. That can be particularly valuable in distilling out the truth from the barrage of conflicting ideas and opinions that we constantly face in our interactions with people and the media. So, in the most literal sense, this is a guide to life, one that comes from the source of all our knowledge and wisdom: nature.

Indeed, it is true that from the earliest history of human knowledge, our primary source of wisdom has always been nature—even our essential spirituality ultimately derives from questions about our place in this universe. So what we have done here is simply delve into our updated knowledge base of nature and the universe to distill bits of meaningful wisdom to apply to our lives. It is an ancient process but built upon modern knowledge.

INDEX

Physics terms

Proper Names

Life and Social terms

Books, Movies, etc

Aphorisms